William J. Deutsch

GROUNDWATER GEOCHEMISTRY
Fundamentals and Applications to Contamination

LEWIS PUBLISHERS
Boca Raton London New York Washington, D.C.

Library of Congress Cataloging-in-Publication Data

Deutsch, William J.
Groundwater Geochemistry: fundamentals and applications to contamination /
        William J. Deutsch
            p.   cm.
        Includes bibliographical references and index.
        ISBN 0-87371-308-7 (alk. paper)
        1. Water chemistry. 2. Groundwater. 3. Groundwater—Pollution. I. Title.
GB855.D48 1997
551.49—dc21   97-11261
                                                                                          CIP

Visit the CRC Web site at www.crcpress.com

© 1997 by CRC Press LLC
Lewis Publishers is an imprint of CRC Press LLC

No claim to original U.S. Government works
International Standard Book Number 0-87371-308-7
Library of Congress Card Number 97-11261
Printed in the United States of America   7 8 9 0
Printed on acid-free paper

DEDICATION

For Marie Pavish
Elizabeth, Emily, and Annika

AUTHOR

William J. Deutsch attended the University of Washington in Seattle, where he earned Bachelor of Science degrees in Geological Sciences and Oceanography in 1970. For the next three and a half years, he was an oceanographic officer in the U.S. Navy stationed onshore in Bermuda and onboard a research ship homeported in Japan. Following military duty, Deutsch studied earth sciences at Scripps Institute of Oceanography and then returned to the University of Washington, where he earned a Master of Science degree in Geological Sciences in 1978. Deutsch's primary areas of interest in graduate school were soil processes, chemical weathering and water–rock interactions.

Deutsch joined Battelle's Pacific Northwest Laboratories in Richland, Washington, as a geochemist in 1979. Early work at Battelle centered on developing predictive techniques for estimating the effect of changing environmental conditions on groundwater chemistry, and this lead to his introduction to geochemical modeling. While at Battelle, Deutsch managed a variety of projects dealing with groundwater contamination, aquifer restoration and geochemical modeling. These projects involved all phases of investigations from field studies, to laboratory research and computer simulations of natural systems.

In 1985 Deutsch joined Jacobs Engineering as the senior geochemist on the Uranium Mill Tailings Remedial Action (UMTRA) project managed from Albuquerque, New Mexico. In this role and his subsequent position as Acting Manager of Hydrology for the UMTRA project, he was responsible for hydrogeological investigations and the design of remedial action plans for water resource protection at the 24 UMTRA sites.

Deutsch joined Woodward-Clyde in 1987. Over the past 10 years he has managed projects and provided geochemistry support on the full range of environmental projects dealing with both organic and inorganic contaminants at industrial sites, military bases, commercial outlets, etc. He has been involved in all phases of remediation work from investigation, design, implementation, maintenance and monitoring. For the past 10 years he has taught professional development courses on groundwater geochemistry and geochemical modeling for the National Ground Water Association. He is an NGWA Certified Ground Water Professional.

He is a member of the American Chemical Society, National Ground Water Association, and the Society of American Military Engineers.

PREFACE

The discovery of widespread groundwater contamination in the last few decades and growing awareness of the importance of this vulnerable resource has lead to extensive efforts throughout the world to protect clean groundwater and remediate contaminated aquifers. Aquifer protection requires effective groundwater monitoring, and remediation requires adequate site characterization to identify the sources, levels, and mobilities of contaminating substances. Groundwater geochemistry is a necessary component of the evaluation of vadose zone and aquifer systems. Geochemistry considers the chemical interactions that occur between water, solids, and gases in the subsurface. These reactions change the composition and concentration of naturally occurring and contaminant compounds, thereby impacting contaminant mobility and susceptibility to remediation. Effective monitoring and remediation programs require consideration of the geochemistry of the system.

This book has been developed to provide readers of all backgrounds the necessary information to gain a better understanding of geochemical processes and their application to groundwater systems. To widen the appreciation of geochemistry as a part of groundwater investigations as much as possible, we have not presumed that the reader has an extensive technical background in aqueous chemistry, geochemistry, or geochemical modeling. Our only presumption is that the reader shares our enthusiasm for understanding the natural world and would like to focus that energy on obtaining a better understanding of groundwater geochemistry.

Part I of this book provides basic information on geochemical processes for environmental scientists and engineers with little training or experience in geochemistry. By understanding these fundamental geochemical processes, the investigator will gain insight into why groundwater concentrations change along the flow path and will be able to better explain temporal and spatial trends in data. Also described in Part I are the methods and tools that can be used to understand and simulate the geochemical processes. Part II covers the general application of geochemistry to contaminant mobility, the design of remediation systems, the development of effective sampling programs, and the modeling of geochemical interactions. Part III describes specific applications of geochemistry to sites contaminated by landfills, acidic materials, metals, and organic compounds.

The material presented in this book has been substantially derived from short courses on groundwater geochemistry and geochemical modeling given by the author through various professional organizations over the past 10 years. The courses are focused at individuals with all levels of education and training and are designed to give professional scientists and engineers practical information on the fundamentals and applications of geochemistry to groundwater. This book has been written as a supplement to the courses and to disseminate this information to a wider audience.

ACKNOWLEDGMENTS

This book would never have seen the light of day without the encouragement and support of my family, friends, and colleagues. My parents started it all a few years ago, and my wife and children stuck by me through thick and thin when the words flowed easily and when there was turbulence. I would like to thank Pat Longmire, Rick Meyerhein, Stan Peterson, and Van Ekambaram for all they have taught me during the groundwater chemistry short courses that we have presented over the years. Stan also reviewed parts of the book and helped me settle on the outline. Bob Moran provided valuable input on the geochemistry of acid mine drainage and also reviewed parts of the book. I owe a tremendous debt of gratitude to Kay Thompson and Jane Werthheimer for taking my crude drawings and turning them into the works of art provided in the figures.

I would also like to acknowledge the support provided by Woodward-Clyde during my sabbatical and, in particular, the faith that Jim Miller showed in me. Finally, I appreciate the essential contributions of the staff at CRC Press including Publisher Joel Stein and Project Editor Albert Starkweather, as well as the copy editor, Elizabeth Mahoney.

TABLE OF CONTENTS

PART III: APPLICATIONS OF GEOCHEMISTRY TO SPECIFIC TYPES OF CONTAMINANTS AND CONTAMINATED ENVIRONMENTS

PART I.
GROUNDWATER GEOCHEMISTRY FUNDAMENTALS

Water present below the land surface takes on some of the characteristics of that environment. Rainfall and snowmelt percolating through the soil zone and unsaturated material chemically react with the gases, minerals, and organic compounds that occur naturally in the subsurface. These reactions continue below the water table as the water flows through the aquifer. The result is that the characteristics and composition of the water evolve as it flows through the ground in response to the types of solid and gas phases that the solution encounters and the geochemical reactions that occur between these phases. Part I of this book describes the fundamentals of groundwater geochemistry. Chapter 1 discusses characterization of the water/rock system, and Chapters 2 and 3 discuss solution/gas and water/rock reaction processes. The development of geochemical models to simulate the natural system is described in Chapter 4, while Chapter 5 provides a discussion of the computer codes commonly used to model groundwater geochemical reactions.

1 THE GROUNDWATER GEOCHEMICAL SYSTEM

Groundwater supplies drinking water to over half the population of the United States and provides an even greater percentage of the amount of water used for irrigation and industrial purposes.[1] Thirty years ago the main use of geochemistry to groundwater was as an aid in differentiating water types in aquifers to help identify and quantify water resources. Today, geochemistry plays a much larger role in groundwater studies because of its importance in characterizing the natural system, understanding contaminant migration, and designing remediation programs. Neglecting the geochemistry component of groundwater contamination may tremendously oversimplify the situation and lead to erroneous conclusions regarding the impact of contamination on the environment and the amount of effort required to remediate the aquifer.

The area of geochemistry that we are interested in is low temperature, aqueous geochemistry that focuses on water/rock interactions in the unsaturated zone and below the water table in the aquifer(s) present at a site. These water/rock interactions are the geochemical processes that soil scientists have studied for decades to optimize plant growth and that groundwater scientists/engineers are now using to better understand the chemical facet of subsurface conditions. Along with the physical and biological processes active in the subsurface, geochemistry plays a major role in controlling groundwater composition and the movement of dissolved constituents.

The subsurface can be thought of as a dynamic geochemical system consisting of (1) solid phases (minerals, amorphous [noncrystalline] solids, and organic matter), (2) a soil gas phase, and (3) an aqueous solution phase (water with its dissolved constituents). Each phase is a separate physical entity with more or less uniform chemical composition. There is commonly a single gas phase and a single solution phase, although two or more separate solution phases (such as surface and groundwater) may mix to produce a new solution phase. Each mineral type or other solid material with the same composition and properties is considered a separate phase. A system is simply an assemblage of these solid, gas, and solution phases under site-specific temperature and pressure conditions. The beaker model shown in Figure 1-1 represents a simple geochemical system. The five phases comprising this system are solution, gas, organic matter (CH_2O), biotite, and calcite. Water is in contact with the gas phase and the various inorganic/organic solid phases. Some of the constituents of the solution phase may have been present as dissolved in the water when it was added to the beaker, whereas others may have dissolved into the water as a result of reactions in the beaker between the phases.

In the natural geochemical system, fresh recharge water with few dissolved constituents and low concentrations contacts the subsurface material and interacts with the other phases of the system. Chemical reactions occur because the composition of the recharge water is not in equilibrium with the solid phases or the soil gases. Disequilibrium drives the reactions that dissolve gases and minerals into the water and changes the solution composition. New minerals may precipitate from water as the solution composition becomes saturated with the constituents of secondary minerals. These secondary minerals are weathering products of the dissolution of the primary minerals.

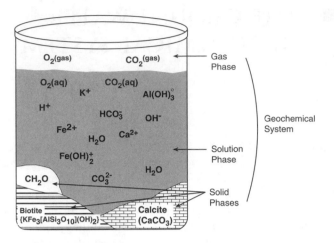

Figure 1-1   Beaker model of geochemical system.

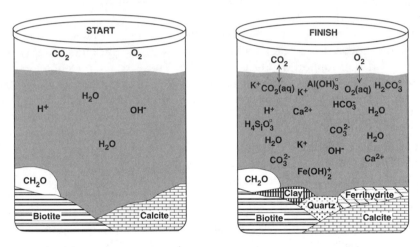

Figure 1-2   Water/rock/gas interactions.

The reaction process can be illustrated by comparing the two beakers shown in Figure 1-2. In the first beaker, fresh water without any dissolved constituents other than the components of water comes into contact with atmospheric air (or soil gas in the subsurface) and minerals and solid organic matter. Because few of the components of the air or solids are initially present in the pure water, disequilibrium exists and dissolution reactions occur that provide those components to the water. Reactions of the following types produce the system shown in the second beaker.

$$CaCO_3(calcite) + H^+ \rightarrow Ca^{2+} + HCO_3^- \tag{1-1}$$

$$2KFe_3AlSi_3O_{10}(OH)_2(biotite) + 14H^+ + H_2O \rightarrow$$
$$2K^+ + 6Fe^{2+} + 4H_4SiO_4^0 + Al_2Si_2O_5(OH)_4(kaolinite) \tag{1-2}$$

$$CH_2O(organic\ matter) + O_2(aq) \rightarrow CO_2(gas) + H_2O \tag{1-3}$$

The dissolved solutes in the water are derived from the dissolution or leaching of the solid phases and from the dissolution of gases from the air or production of gases by chemical reaction (e.g., carbon dioxide from the oxidation of organic matter). These reactions continue as long as disequilibrium exists between the phases. Solution concentrations may stabilize if the dissolved concentrations of the components of the solids and/or gases achieve chemical equilibrium. For instance, if we add calcite ($CaCO_3$) to pure water, the solid calcite will dissolve until the concentration of its components (calcium and carbonate) in the solution have reached a level in equilibrium with the solid mineral. If the conditions of the system do not change, then the dissolved concentrations of calcium and carbonate will be fixed at the values in equilibrium with calcite. To understand the natural geochemical system, we need to first characterize the phases that comprise the system. In this chapter we will introduce the solution and solid phases important in the study of groundwater. In subsequent chapters we will discuss the gas phase and describe in detail the reactions that occur between these phases.

1.1. GROUNDWATER SOLUTION

The solution phase in the subsurface is the primary medium of exchange and transport. At the water table and above in the unsaturated zone gasses dissolve into the water and can be transported away from their point of origin. Along the entire flow path from the earth's surface through the aquifer, components of the solid phases dissolve into and/or precipitate from the groundwater, which then facilitates or retards the movement of dissolved components through the system. This section provides information on characterizing the solution component of the system and describes the more important chemical reactions that occur in the solution phase.

Definitions and Concentration Units

Groundwater is an aqueous solution in the subsurface. A solution can be defined as a solvent (in this case water) with dissolved inorganic or organic constituents (solutes). Figure 1-3 shows the relationship between the solvent, solutes, and a solution. The concentration of a solute in a solution is commonly reported in units of milligrams per liter (mg/L) or parts per million (ppm). As depicted in Figure 1-4, units of milligrams per liter pertain to the mass (milligrams) of a solute per liter (volume) of a solution. Parts per million are defined as mass of solute per kilogram (mass) of solvent. Units of mg/L are given in terms of a liter of solution and ppm relate to a kilogram of solvent. Even though one liter of pure water has a mass of one kilogram at 3.89°C, these two measures of the solvent are not equivalent at other temperatures. The addition of solutes to the kilogram of water will raise its volume, further differentiating the two concentration units. The following conversion can be used to change units if data from water analyses are provided in mixed units:

$$\frac{mg/L}{ppm} = \text{solution density } (g/cc) - TDS \, (g/cc)$$

For example, if solution density equals 1.008 g/cc and the total dissolved solids (TDS) level of the solution is 10,000 mg/L (0.01 g/cc), the ratio of mg/L to ppm is 0.998 g/cc (that is, 1 ppm = 0.998 mg/L). In this case the difference in the concentration units is 0.2%. At solution concentrations higher than about 7,000 mg/L, it is recommended that density corrections be made in converting concentration units.[2] At lower concentration levels, the two units are often used interchangeably.

For the purpose of understanding geochemical reactions in the environment, it is necessary to convert typical laboratory concentration units of milligrams per liter or parts per million to a scale that employs units of moles per liter or moles per kilogram. A mole refers to the number of atoms or molecules of a solute in the solution. One mole corresponds to Avogadro's Number (6.022×10^{23}) of atoms or molecules of the constituent and has a

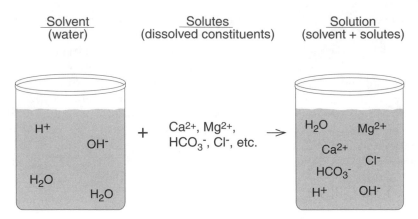

Figure 1-3   Components of a solution.

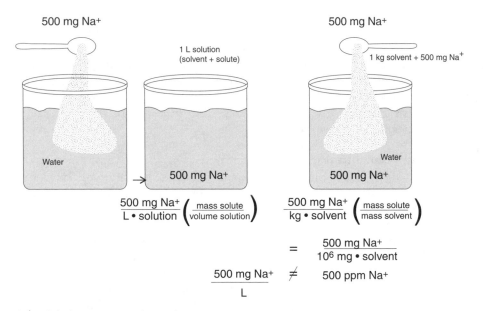

Figure 1-4   Solution concentration units.

mass equal to the atomic or molecular weight of the constituent. For example, the atomic weight of carbon is 12; therefore 12 grams of carbon (one mole) contain 6.022×10^{23} atoms of carbon. The molecular weight of carbonate (CO_3^{2-}) is 60; therefore, 60 grams of carbonate contain 6.022×10^{23} molecules of carbonate.

The concentration units applicable to the mole scale are molarity (M) and molality (m). Molarity is the number of moles of a solute in one liter of solution, and molality is the number of moles of a solute in one kilogram of solvent. Molarity is analogous to milligrams per liter, and molality is analogous to parts per million. At higher concentration levels, density corrections should be made in converting concentration units using the mole scale as was necessary using the mass scale.

The reason that the mole scale is appropriate for understanding the geochemistry of a system is that chemical reactions and the thermodynamic data (e.g., equilibrium constants) that have been determined for these reactions are in terms of molarity or molality. We can write a reaction for the formation of the mineral calcite as follows:

$$Ca^{2+} + CO_3^{2-} \rightarrow CaCO_3 \quad K = 10^{-8.4} \left(25°C\right) \tag{1-4}$$

The stoichiometry of this reaction shows that one atom of calcium will combine with one molecule of carbonate to form one unit of solid calcite. Because the stoichiometry corresponds to atoms and molecules and not mass of the reactants or products, the appropriate units for characterizing the reaction are moles per liter/kilogram and not mass. The equilibrium constant (K) for the reaction is reported in terms of mole concentration units.

Groundwater Solutes

Water is a very strong solvent capable of dissolving most solids to some degree. Of the many solutes found in groundwater only a relatively few are present at concentrations greater than 1 mg/L under typical natural conditions. These are generally called the major ions, and they consist of the cations calcium (Ca^{2+}), magnesium (Mg^{2+}), sodium (Na^+), and potassium (K^+) and the anions bicarbonate/carbonate (HCO_3^-/CO_3^{2-}), sulfate (SO_4^{2-}), chloride (Cl^-), and nitrate (NO_3^-). An important nonionic constituent in typical groundwater is silicon (Si), which generally occurs as the uncharged dissolved species $H_4SiO_4^0$. The potential presence of any other element from the Periodic Table in groundwater is limited only by contact of the solution with a solid or gas containing the element; however, most other elements and compounds are found dissolved in groundwater at levels less than 1 mg/L. The more common minor and trace elements are listed in Table 1-1. The concentrations of the major ions in groundwater and the concentrations of the other dissolved species are a function of the availability of the constituent to the system and the solubility of solids that may limit solution concentration. Processes limiting solution concentration levels are discussed in detail in Section 1.4.

TABLE 1-1.

TYPICAL GROUNDWATER DISSOLVED CONSTITUENTS	
Major ions	**Minor and trace constituents**
Ca^{2+}, Mg^{2+}, Na^{2+}, K^+,	Fe, Mn, Al, Ba, Cd,
HCO_3^-/CO_3^-, SO_4^-,	Cr, Pb, Zn, As, Se,
Cl^-, NO_3^-, $H_4SiO_4^0$	P, F, Br, I, B,
	NH_4^+, Co, Cu, Pb, Hg

Because the common major solutes in groundwater are positively and negatively charged species and the groundwater solution must be electrically balanced, laboratory data can be evaluated by calculating whether the measured concentrations provide an electrically neutral solution. This is done by converting measured concentrations in mg/L (or ppm) to an electrical equivalent unit for each major ionic species using the following formula:

$$\text{Equivalent/liter} = \text{Concentration (mg/L)} \times \frac{1 \text{ mole}}{\text{atomic or molecular wt.}} \times \left| \frac{\text{\# of equivalents}}{\text{mole}} \right|$$

In this equation, the number of equivalents per mole is the valence (charge) on the dissolved species. As an example, calcium has a valence of +2 and an atomic weight of 40.08 g. If the laboratory reported the concentration of calcium (Ca^{2+}) in a water sample as 92 mg/L, then the number of equivalents per liter attributed to this concentration of dissolved calcium is as follows:

$$92 \text{ mg/L} \times 1 \text{ mole}/40.08 \text{ g} \times 2 \text{ equiv/mole} = 4.6 \times 10^{-3} \text{equiv/L}$$

Because typical groundwater concentrations correspond to equivalents that are on the order of 10^{-3}, it is convenient to report results in milliequivalents, where 1000 milliequivalents (meq) equals 1 equivalent. In the calcium example, 4.6×10^{-3} equiv/L equals 4.6 meq/L.

Table 1-2 shows the data and steps used to convert a typical groundwater analysis to electrical equivalents for the major ions. The electrical balance of the solution can be calculated by comparing the sum of the equivalents due to the cations with the sum of the equivalents due to the anions. The following formula is used to make this comparison:

$$\text{Cation/anion balance} = \frac{\Sigma(\text{cations, meq/L}) - \Sigma(\text{anions, meq/L})}{\Sigma(\text{cations, meq/L}) + \Sigma(\text{anions, meq/L})} \times 100\% \qquad (1\text{-}5)$$

TABLE 1-2.

EXAMPLE OF ELECTRICAL EQUIVALENTS CALCULATION PROCEDURE

Const.	Measured conc. (mg/L)	Atomic weight (g)	Molarity (mmol/L)	Valence	meq/L
Ca^{2+}	92.0	40.08	2.30	+2	4.60
Mg^{2+}	34.0	24.31	1.40	+2	2.80
Na^+	8.2	23.0	0.36	+1	0.36
K^+	1.4	39.1	0.04	+1	0.04
Fe(III)	0.09	55.8	0.002	+3	0.006
HCO_3^-	339.0	61.0	5.56	−1	5.56
SO_4^{2-}	84.0	96.0	0.88	−2	1.7
Cl^-	9.6	35.5	0.27	−1	0.27
NO_3^-	13.0	62.0	0.21	−1	0.21

Note: meq/L = milliequivalents/liter.

$$Ca^{2+}: \frac{92 \text{ mg}}{L} \times \frac{1 \text{ g}}{1000 \text{ mg}} \times \frac{1 \text{ mole}}{40.08 \text{ g}} \times \frac{10^3 \text{ millimoles}}{\text{mole}} \times \frac{2 \text{ milliequiv}}{\text{millimole}} = \frac{4.6 \text{ meq}}{L}$$

The cation/anion balance for the data in Table 1-2 is +0.5%. A positive number means that either there are excess cations or insufficient anions in the analysis, whereas a negative balance corresponds to excess anions or insufficient cations. A reasonable balance for routine analyses is generally considered to be less than 5%. Several possible reasons, alone or in concert, can create an electrical imbalance in the reported composition data:

1. The design of the sampling program neglected a major dissolved species.

2. Laboratory error.

3. Using unfiltered water samples that contain particulate matter that dissolves in the sample when acid is added for preservation purposes.

4. The precipitation of a mineral in the sample container that removes the constituents of the mineral from the water.

5. In certain cases the dissolved species of the element or compound may not correspond to the typical species used in making the ion balance calculation.

An example of a situation when the last case can be important is the distribution of inorganic carbon species. At normal groundwater pH values of 6.5 to 8.5 the dominant inorganic carbon species in groundwater is bicarbonate (HCO_3^-) with a valence of one. Because it is usually the dominant inorganic carbon species, bicarbonate is normally used to calculate the amount of equivalents in solution contributed by inorganic carbon. At pH values greater than 9, the inorganic carbon species carbonate (CO_3^{2-}) with a valence of two becomes a progressively more important dissolved constituent. If the pH of the water is greater than 9, the distribution of inorganic carbon species between bicarbonate and carbonate must be calculated as part of the ion balance calculation.

Fritz[3] compared charge balances for over one thousand water analyses reported in the literature. He found an average imbalance of 3.99% ± 6.56 (1σ). He most commonly found imbalances on the positive side (excess cations or insufficient anions). The primary reason he attributed for a positive balance was the determination of alkalinity (a measure of the inorganic carbon concentration) in the laboratory on unacidified samples. Measuring alkalinity in this manner may not account for the carbonate that precipitated in the bottom of the sample bottle, thereby lowering the actual anion concentration. The primary reason he attributed for a negative balance (too high a level of anions or not enough cations) was failure to filter the alkalinity samples collected from an aquifer containing carbonate minerals. This allowed particulate calcite in the water sample to be titrated during the alkalinity determination resulting in excess reported dissolved inorganic carbon. Sampling protocol and alkalinity measurements are discussed further below in the section on alkalinity and in Section 9.5. Fritz also pointed out the difficulty in achieving a good ion balance in dilute groundwaters of low TDS because small errors become more pronounced when the denominator (sum of the cations and anions) is small in the ion balance equation.

To use the cation/anion balance equation to check a sampling program or evaluate a laboratory, it is necessary that the analyses be for the dissolved constituents only. Groundwater may contain suspended particulate matter, as well as dissolved constituents. The particulates may be inorganic minerals or amorphic solids, as well as organic matter. Because they are not dissolved in the water, they are not part of the charge balance for the solution, therefore they must be filtered from the water before preserving the water with a substance, for example, an acid, that might dissolve some or all of the particulate matter. Groundwater is typically filtered with a 0.45-micrometer pore size filter to remove suspended material prior to analysis. However, it has been shown[4] that smaller pore sizes on the order of 0.1 micrometer or less may be necessary in some situations to remove suspended particles of metal oxides (particularly those of Fe, Al, and Mn) and perhaps clay minerals that are smaller than 0.45 micrometer. From a practical standpoint, water that contains even small amounts of particulates is difficult to force through filter pore sizes much smaller than 0.45 micrometer.

In addition to providing correct data for calculating ion balances, it is also necessary to measure the concentrations of dissolved species to calculate mineral equilibria relationships. As will be discussed in Section 3.2, the dissolved levels of groundwater constituents are compared with expected levels in equilibrium with minerals to evaluate whether or not the solution is in equilibrium with the mineral. Total concentrations in solution (dissolved plus particulate) cannot be used in the equilibrium calculations if the particulate level is a significant portion of the total. The total concentration of a contaminant is an important measure of the potential risk to human health or the environment from the water and may be a useful measurement to make for a water sample in addition to dissolved concentration levels.

Groundwater Types

The major ion composition of groundwater is used to classify groundwater into various types based on the dominant cations and anions. For example, if calcium and bicarbonate are the dominant cation and anion, then the groundwater would be a Ca–HCO_3 type. The composition of the dominant ions can be displayed graphically by several methods (see Hem[2]) with one of the more useful summary presentations being the trilinear or Piper diagram.[5] On this diagram the relative concentrations of the major ions in percent Meq/L are plotted on cation and anion triangles, and then the locations are projected to a point on a quadrilateral representing both cation and anions. Figure 1-5 shows the basic diagrams used to plot water analyses using the trilinear method; also shown are the fields for different water types.

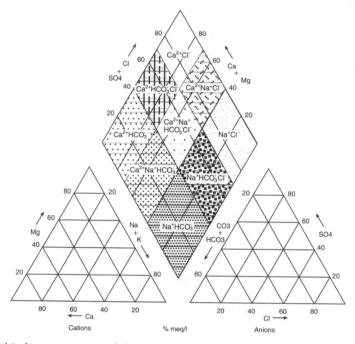

Figure 1-5   Graphical representation of data — trilinear diagrams.

White et al.[6] provide an extensive compilation of groundwater analyses from a wide variety of aquifer host rocks. The aquifer material includes igneous rocks (granites and basalts), sedimentary rocks (sandstone, shale, limestone, and dolomite), a metamorphic quartzite, and unconsolidated sand and gravel. Several water analyses have been averaged for five to ten separate samples from each aquifer type and are plotted in Figure 1-6. It is somewhat surprising that all but one of the aquifer rock types has bicarbonate as the dominant anion. The only exception is the water (#9) from oil and gas field regions in which chloride is the dominant anion. In aquifers without primary carbonate minerals, such as granite, basalt, and quartzite aquifers, the source of bicarbonate is carbon dioxide gas in atmospheric air and soil vapor and low concentrations of carbonate minerals present as weathering products. As shown in Figure 1-6, calcium is the dominant cation in several of the aquifers (limestone, dolomite, quartzite, and sand/gravel) and is present in high concentration in those dominated by another cation such as sodium (silicic igneous rock, sandstone, shale-claystone) and magnesium (gabbro-basalt). The source of the cations in groundwater is the weathering of the predominantly silicate and carbonate minerals in the aquifer host rock. Hem[2] provides several tables of groundwater analyses grouped by major constituent and shows that some distinctive groundwaters may have iron, manganese, or aluminum present in relatively high concentration.

Hydrogen Ion Activity (pH)
The total hydrogen concentration is not normally determined for a solution; instead the activity of the free, uncomplexed hydrogen ion is measured. Because of the reactivity of the hydrogen ion, its activity (or effective concentration) in groundwater is an especially

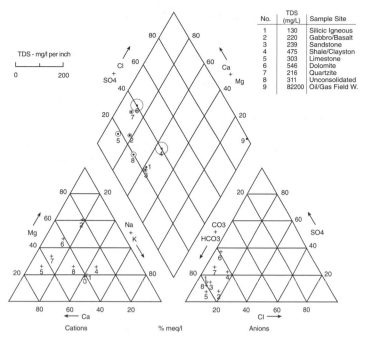

No.	TDS (mg/L)	Sample Site
1	130	Silicic Igneous
2	220	Gabbro/Basalt
3	239	Sandstone
4	475	Shale/Clayston
5	303	Limestone
6	546	Dolomite
7	216	Quartzite
8	311	Unconsolidated
9	82200	Oil/Gas Field W.

Figure 1-6   Survey of groundwater types.

important parameter. The activity of hydrogen is measured as the pH of the water, which is defined as follows:

$$pH = -\log_{10}\left(a_H^+\right) \tag{1-6}$$

where a_H^+ is the activity of the hydrogen ion (moles/kg).

The hydrogen ion activity is a master variable of the groundwater system because the hydrogen ion participates in most of the chemical reactions that affect water composition. For example,

Mineral dissolution/precipitation: $CaCO_3 + H^+ = Ca^{2+} + HCO_3^-$

Aqueous complexation: $Fe^{3+} + H_2O = Fe(OH)^{2+} + H^+$

Adsorption/desorption: $\equiv FeOH + Cu^+ = \equiv FeOCu + H^+$

The measured pH of a system does not by itself provide any information on the capacity of the system to maintain (buffer) that pH as an acid or base is added to the system. The buffering capacity of the system is extremely important because of this ability to maintain the pH of the system. Because inorganic carbon species are often the dominant anion in groundwater systems and they can take up or release hydrogen ions as part of their speciation reactions, they provide much of the buffering capacity in natural water. Inorganic carbon species buffer the pH by the following reactions:

$$H_2CO_3 \leftrightarrow HCO_3^- + H^+ \quad K = 10^{-6.4} \tag{1-7}$$

$$HCO_3^- \leftrightarrow CO_3^{2-} + H^+ \quad K = 10^{-10.3} \tag{1-8}$$

As an acid is added to a solution containing inorganic carbon, some of the carbonate (CO_3^{2-}) is consumed to form bicarbonate (HCO_3^-) and some of the bicarbonate becomes carbonic acid (H_2CO_3). These reactions tie up the added hydrogen ion, lowering its effect on the solution pH, which is a measure of only the free uncomplexed H^+. For example, if 0.5 ml of concentrated nitric acid (16.8 moles HNO_3/L) were added to one liter of pure water (pH = 7) the final pH of the solution would be 2.1. If this same amount of acid were added to a groundwater sample with an initial pH of 6.4 and equal dissolved concentrations of 2×10^{-3} mol/L (about 120 mg/l) H_2CO_3 and HCO_3^-, the final pH would be 6.0.

If a base were added to the solution, the carbonic acid would be converted to bicarbonate and some of the bicarbonate would convert to carbonate, thereby consuming some of the added hyroxyl ions (OH^-) according to the following reactions:

$$H_2CO_3 + OH^- \leftrightarrow HCO_3^- + H_2O \quad K = 10^{-7.6} \tag{1-9}$$

$$HCO_3^- + OH^- \leftrightarrow CO_3^{2-} + H_2O \quad K = 10^{-3.7} \tag{1-10}$$

The strongest buffering due to these solution species occurs when the dissolved inorganic carbon concentration is high and the two carbon components of the reaction are present in equal concentration, which occurs at pH values of 6.4 and 10.3 according to the equilibrium reactions (Equations 1-7 and 1-8) for the carbonate species. Buffering occurs at other pHs within ±1 pH units of these values, but not as strongly.

Alkalinity/Acidity

Alkalinity is a measure of the total acid-neutralizing capacity of the water, and acidity is the base-neutralizing capacity of the water. Whereas buffering reactions tend to maintain a pH value at or near a specific value as an acid or base is added to the water, alkalinity and acidity represent the cumulative reactions that consume hydrogen as acid is added to the solution (alkalinity) or that release hydrogen as a base is added to the solution (acidity). Alkalinity is the acid-neutralizing capacity (ANC), and acidity is the base-neutralizing capacity (BNC) of the solution. The alkalinity of a solution is determined by titrating a water sample with an acid (commonly H_2SO_4) to an endpoint pH of about 4.5. The acidity of a solution is determined by titrating a water sample with a base (commonly NaOH) to an endpoint pH of about 8.3. The handbook of Standard Methods[7] provides the procedures for alkalinity (method 2320) and acidity (method 2310) titrations.

Alkalinity is much more commonly determined for a water sample than acidity because most of the alkalinity in a groundwater is due to the amount of inorganic carbon ions present in solution, therefore the alkalinity measurement can be used to determine the bicarbonate and carbonate concentration of the groundwater. This is an important measurement because bicarbonate is often the dominant anion in shallow groundwater. The definition of alkalinity actually includes all species dissolved in the water that can potentially neutralize acid; therefore a complete representation of alkalinity is as follows:

$$\text{Alkalinity (meq/L)} = m_{HCO_3^-} + 2m_{CO_3^{2-}} + m_{H_3SiO_4^-} + m_{\text{organic ions}} + \dots + m_{OH^-} - m_{H^+} \tag{1-11}$$

This equation shows that alkalinity (in milliequivalents per liter) is the sum of the concentrations (m, moles/kg) of all of the dissolved species that might accept (consume) a hydrogen ion during titration. If a species can accept more than one hydrogen ion, then its concentration is multiplied by the appropriate factor. The factor of two in Equation 1-11 for carbonate is present because CO_3^{2-} can accept two hydrogen ions to form carbonic acid, H_2CO_3. When the alkalinity measurement is used to determine the bicarbonate and carbonate species in solution, the assumption is made that all the other possible titratable species have negligible concentrations. This may not be a good assumption at pH values greater than 9 where the silicon species $H_3SiO_4^-$ can be important or in areas with high organic contamination such as near landfills where titratable organic anions may be present in relatively high concentration.[8,9] In these cases the noncarbonate alkalinity must be subtracted from the total alkalinity before inorganic carbon concentration can be calculated (see Section 9.5).

Alkalinity is normally reported by the laboratory in units of mg/L as $CaCO_3$. Because each carbonate ion is capable of neutralizing two hydrogen ions, a factor of two must be used to convert these units to meq/L per the following equation:

$$\text{Alkalinity} \left(\text{meq/L}\right) = \text{Alkalinity} \left(\text{mg/L as CaCO}_3\right) \times 2 \text{ meq/mmole} \times 1 \text{ mmole/100 mg} \quad (1\text{-}12)$$

If all of the alkalinity in the sample is a result of inorganic carbon species, the distribution of these species between bicarbonate and carbonate can be calculated using the equilibrium constants for the speciation reaction (Eq. 1-8) and the measured pH of the solution with the following equations:

$$\text{HCO}_3^- \left(\text{mg/L}\right) = \frac{\text{Alkalinity} \left(\text{mg/L CaCO}_3\right)}{\left(1 + \dfrac{2 \times 10^{-10.3}}{10^{-\text{pH}}}\right) \times 50} \times 61 \quad (1\text{-}13)$$

$$\text{CO}_3^{2-} \left(\text{mg/L}\right) = \frac{\text{Alkalinity} \left(\text{mg/L CaCO}_3\right)}{\left(2 + \dfrac{\left(10^{-\text{pH}}\right)}{10^{-10.3}}\right) \times 50} \times 60 \quad (1\text{-}14)$$

Example 1-1 shows a typical calculation for a groundwater sample. As stated above, if other dissolved species provide alkalinity to the water, then their concentrations must be determined separately and subtracted from the alkalinity value prior to the bicarbonate/carbonate calculation.

Example 1-1. Converting measured alkalinity in terms of $CaCO_3$ (mg/L) to concentrations of bicarbonate and carbonate.

Alkalinity measured at 100 mg/L $CaCO_3$; pH = 8.5

$$\text{HCO}_3^- \left(\text{mg/L}\right) = \frac{\text{Alkalinity} \left(\text{mg/L CaCO}_3\right)}{1 + \dfrac{2 \times 10^{-10.3}}{\left[a_{H^+}\right]} \times 50} \times 61$$

$$= \frac{100}{1 + \dfrac{2 \times 10^{-10.3}}{\left[10^{-8.5}\right]} \times 50} \times 61 = 118 \text{ mg/L HCO}_3^-$$

$$CO_3^{2-} \, (mg/L) = \frac{Alkalinity \, (mg/L \, CaCO_3)}{2 + \dfrac{[a_{H^+}]}{10^{-10.3}} \times 50} \times 60$$

$$= \frac{100}{2 + \dfrac{[10^{-8.5}]}{10^{-10.3}} \times 50} \times 60 = 1.8 \, mg/L \, CO_3^{2-}$$

In the subsurface the acid-neutralizing capacity of the system must consider not just the alkalinity of the groundwater but the minerals in the rock that may also react with acid added to the system and neutralize some of the hydrogen ion. Most weathering reactions consume hydrogen, such the weathering of albite to the clay mineral kaolinite:

$$2NaAlSi_3O_8(albite) + 2H^+ + 9H_2O \rightarrow Al_2Si_2O_5(OH)_5(kaolinite) + 2Na^+ + 4H_4SiO_4 \quad (1\text{-}15)$$

However, the weathering of silicate minerals is a relatively slow process compared with the reaction of carbonate minerals under acidic conditions. In most cases the acid-neutralizing capacity of the solid phase of an aquifer is due to its carbonate mineral content (calcite, dolomite, siderite, etc.). Section 12.1 contains additional discussion on acid neutralizations in natural systems.

The acidity of water is a function of the concentration of dissolved species that can contribute a hydrogen ion to solution to consume hydroxyl ions added during the acidity titration. The acidity can be defined similarly to alkalinity as follows:

$$Acidity \, (meq/L) = 2m_{H_2CO_3} + m_{HCO_3^-} + m_{organic \, acids} + 2m_{H_2S} + m_{HS^-} + m_{H^+} - m_{OH^-} \quad (1\text{-}16)$$

Acidity is not commonly measured for water samples because it cannot be used to determine the carbonate ion concentration, which is obtained with an alkalinity measurement. However, it can be an important measurement to make if it is necessary to neutralize acidic water. The acidity measurement will quantify all those processes in the solution that will inhibit neutralization, some of which may not be intuitively obvious. For instance, at pH values less than about 4, dissolved metals such as iron and aluminum may be present in groundwater at relatively high concentrations (>1 mg/L). As a base is added to neutralize this water, the metals will hydrolyze and precipitate from solution, thereby consuming some of the added base. This process may be represented by the following reaction:

$$Fe^{3+} + 3OH^- \rightarrow Fe(OH)_3 \quad (1\text{-}17)$$

If iron, or another metal that hydrolyzes in a similar manner, is present at significant concentrations in the groundwater, then its concentration must be added to the acidity equation and the concentration must be multiplied by the appropriate equivalence factor.

In a subsurface system, additional hydrogen ions may also be available to neutralize added base. At low pH there may exist a significant reservoir of exchangeable hydrogen ions on the clay minerals. As the pH of the solution is raised with the addition of the base, cation exchange will provide the hydrogen ions to solution, converting some of the added hydroxyl ions to water.

Redox Potential (Eh)

The Eh, or redox potential, of the system is another master variable like pH. Knowledge of the system Eh is necessary to calculate the distribution of redox-sensitive species between their possible redox states. For example, iron occurs dissolved in groundwater in the ferrous [Fe(II)] and ferric [Fe(III)] redox (valence) states. With a measured Eh for the groundwater, it is possible to calculate from a total dissolved iron concentration [Fe(II) + Fe(III)] the amount of iron that is present in the ferrous and ferric forms, assuming redox equilibrium. Many of the dissolved elements in groundwater occur in more than one redox state (e.g., Mn, N, S, O, Cr), and to calculate mineral equilibrium and other chemical reactions for these elements, it is necessary to have an accurate measure of the redox potential of the system. Unfortunately, it has been shown in many cases[10] that redox equilibrium does not always exist between the various dissolved redox states of an element, and the measured redox potential is not universally applicable to all the redox-sensitive species found in groundwater. For example, the calculated Eh assuming equilibrium with measured dissolved oxygen concentration is generally much higher than the Eh of the groundwater measured with a platinum electrode. The primary reason for redox disequilibrium is the slow rate of many redox reactions. To use a measured redox potential for quantitative purposes, it is necessary to choose those redox couples that are known to be reactive on the time scale of interest (see Section 2.1). Section 9.3 describes methods for measuring the Eh of a water sample.

Just as the pH of a solution may be buffered at a particular value, the Eh may be poised near a certain point by reactions that consume or produce electrons. The major redox-sensitive elements such as iron and manganese present in minerals are generally not found in solution at high concentrations (>1 mg/L) above a pH of about 5. As a consequence, the poising ability of the solution alone is not high. However, in an aquifer system, interactions with reactive, redox-sensitive minerals may poise the system at relatively distinct values. Figure 1-7 shows a pH-Eh diagram for groundwater with the reactive redox-sensitive minerals and their dominant solution species shown. If the minerals are present in the system at even low concentrations (<1%), the redox potential may be poised at the boundaries of the solid/solution species. For instance, under oxidizing conditions MnO_2 is the stable Mn(IV) solid. If a reductant is added to the system, the redox potential will decrease relatively easily until it reaches the stability boundary between MnO_2 and the dissolved manganese species Mn^{2+} (which is drawn on the figure at a concentration of Mn^{2+} at 10^{-6} molar). The redox potential will remain poised close to this boundary as the MnO_2 is dissolved, producing higher levels of dissolved manganese. The degree of poising for the system at this Eh will be a function of the amount of MnO_2 present in the aquifer. Similarly, the Eh may be poised at lower values associated with the $Fe(OH)_3/Fe^{2+}$ and perhaps Fe_2O_3/FeS_2 boundaries.

As with the pH, the measured Eh of the system does not provide information on the redox capacity of the system. Because of the uncertainty as to exactly what is determined when measuring a redox potential, the redox capacity of the system may be a better measure of redox properties.[11,12] The redox capacity can be divided into oxidizing capacity and reducing capacity. Oxidizing capacity in an aquifer is a function of the primary reactions that can consume electrons. In many aquifer systems the oxidizing capacity is due to dissolved oxygen and solid Fe(III) and Mn(IV) phases plus total organic carbon (TOC). If the TOC of the aquifer is assumed to behave as quinone,[11] the reactions providing oxidizing capacity to the system can be written as follows:

$$MnO_2 + 4H^+ + 2e^- = Mn^{2+} + 2H_2O \qquad (1-18)$$

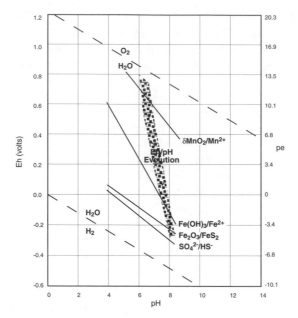

Figure 1-7   Important controls on redox potential in natural water.

$$Fe(OH)_3 + 3H^+ + e^- = Fe^{2+} + 3H_2O \tag{1-19}$$

$$C_6H_4O_2 + 2H^+ + 2e^- = C_6H_6O_2 \tag{1-20}$$

and the oxidation capacity (OXC) of the aquifer solids can be calculated as follows[12]:

$$OXC = \{Fe(III)\} + 2\{Mn(IV)\} + TOC/36 \tag{1-21}$$

where concentrations of the solids in braces are in moles per gram and TOC is measured in mg C/g of solid.

The reducing capacity is commonly due to ferrous iron solids and the TOC. In this case TOC has been assumed to behave as phthalic acid, because it closely resembles humic acids.[11] The reactions providing reducing capacity for the system can then be written as follows:

$$Fe^{2+} + 3H_2O = Fe(OH)_3 + 3H^+ + e^- \tag{1-22}$$

$$C_8H_6O_4 + 12H_2O = 8CO_2 + 30H^+ + 30e^- \tag{1-23}$$

and the reducing capacity (RDC) can be calculated as follows:

$$RDC = \{Fe(II)\} + TOC/3.2 \tag{1-24}$$

where concentrations of the solids in braces are in moles per gram and TOC is measured in mg C/g of solid.

Nikolaidis[12] applied the aquifer redox capacity approach to evaluate whether the natural system could affect the redox state, hence mobility, of the redox-sensitive contaminant chromium.

1.2. CHEMICAL REACTIONS AND THE EQUILIBRIUM CONSTANT

The study of groundwater geochemistry is mainly concerned with the transfer of mass between the various solid, aqueous, and vapor phases that comprise the geochemical system. These changes are described in terms of chemical reactions. Geochemical reactions include mineral dissolution and precipitation, which involves transfer of mass to/from a mineral to the aqueous phase; dissolution of gases from the vapor phase into the aqueous phase or the evolution of gases from the aqueous phase to the vapor phase; and the adsorption/desorption of species between the aqueous phase and the surfaces of the solid phases. Chemical reactions of interest also include transformation reactions that mainly involve solution species. These solution reactions include aqueous complexation and oxidation/reduction reactions. Aqueous complexation occurs when a central molecule becomes associated (complexed) with one or more surrounding molecules to form a complex species in solution (e.g., H_2CO_3, $CaSO_4^0$, $HgCl^+$). Oxidation/reduction reactions involve the transfer of electrons and a change in the valence state of an element (e.g., Fe^{2+} = Fe^{3+} + e^-).

Each chemical reaction has associated with it an equilibrium constant (K). The equilibrium constant is a numerical value that represents the ratio of the activities (effective concentrations) of the participants in the chemical reaction when that reaction has reached chemical equilibrium. At equilibrium, the ratio of the activities will not change and can be represented as a constant (K). For example, calcite dissolution in water can be represented by the following reaction:

$$CaCO_3(s) \rightarrow Ca^{2+} + CO_3^{2-} \tag{1-25}$$

where calcite is a solid (s) phase dissolving into water releasing the dissolved constituents calcium (Ca^{2+}) and carbonate (CO_3^{2-}). (Water can be ignored in writing the reaction because its constituents do not take part in the reaction. Water simply serves as the solvent in which the calcite dissolves.)

If the concentrations of dissolved calcium and carbonate are measured over time as the calcite dissolves, a point will be reached when the level no longer increases. At this point the solution concentrations of the components of calcite are in equilibrium with the mineral, and although the mineral may continue to dissolve in the water, it also precipitates from solution, maintaining a fixed amount of dissolved calcium and carbonate in solution. In equation form, equilibrium can be represented by a reaction that proceeds in both directions as depicted by a double-barbed arrow:

$$CaCO_3(s) \leftrightarrow Ca^{2+} + CO_3^{2-} \tag{1-26}$$

The equilibrium constant for the calcite reaction is equal to the activities of the products of the reaction (calcium and carbonate) multiplied by each other and divided by the activity of the reactant (calcite):

$$K = \frac{\left(a_{Ca^{2+}}\right)\left(a_{CO_3^{2-}}\right)}{a_{CaCO_3}} \times 10^{-8.4} \; (25°C) \tag{1-27}$$

The activities of calcium and carbonate are expressed as moles per kilogram of solution. The activity of the solid phase calcite is effectively constant because it does not matter how much calcite is present at equilibrium. As long as some calcite is present at equilibrium, the product of the activities of dissolved calcium and carbonate will equal $10^{-8.4}$. For this reason the activity of calcite is set equal to 1, and the equilibrium constant expression simplifies to the following:

$$K = \left(a_{Ca^{2+}}\right)\left(a_{CO_3^{2-}}\right) = 10^{-8.4} \; (25°C) \tag{1-28}$$

An alternative method of producing a solution in equilibrium with calcite would be to add the constituents of calcite to the solution until enough are added to reach equilibrium with the mineral. If additional calcium or carbonate are added to solution beyond the equilibrium concentration, calcite precipitates. The precipitation reaction can be written as follows:

$$Ca^{2+} + CO_3^{2-} \rightarrow CaCO_3(s) \tag{1-29}$$

Assuming that the precipitation of calcite is instantaneous, once saturation is reached, the addition of more calcium and carbonate to solution will simply cause more calcite to precipitate and not change the dissolved activities of these two constituents. For the calcite precipitation reaction, the equilibrium constant is written as follows:

$$K = \frac{1}{\left(a_{Ca^{2+}}\right)\left(a_{CO_3^{2-}}\right)} = 10^{8.4} \; (25°C) \tag{1-30}$$

Note that this expression and the numerical value of the equilibrium constant are the inverse of the case of mineral dissolution. The type of chemical reaction under consideration (in this case mineral dissolution versus precipitation) must be known to formulate the equation for the equilibrium constant because K differs according to whether the reaction is written as a dissolution or precipitation process.

The general equilibrium relationship for the reaction

$$aA + bB \leftrightarrow yY + zZ \tag{1-31}$$

is written as follows:

$$K = \frac{\left(a_Y\right)^y \left(a_Z\right)^z}{\left(a_A\right)^a \left(a_B\right)^b} \tag{1-32}$$

where lower case letters represent the stoichiometric coefficients of the reaction.

As mentioned above, when a solid is involved in the reaction, its activity is set equal to one. In addition, the activity of water can generally be set equal to one because, in dilute solutions, the activity of water is essentially constant as small amounts of solids dissolve into or precipitate from the solution. For example, the reaction in which ferrous iron is oxidized and ferric hydroxide precipitates can be written as follows:

$$2Fe^{2+} + 1/2 O_2 + 5H_2O \leftrightarrow 2Fe(OH)_3(s) + 4H^+ \tag{1-33}$$

The complete equilibrium constant relationship for this reaction is:

$$K = \frac{\left(a_{Fe(OH)_3}\right)^2 \left(a_{H^+}\right)^4}{\left(a_{Fe^{2+}}\right)^2 \left(a_{O_2}\right)^{1/2} \left(a_{H_2O}\right)^5} \tag{1-34}$$

but with the simplification of the activities of solids and water set equal to one, the equilibrium constant can be represented as:

$$K = \frac{\left(a_{H^+}\right)^4}{\left(a_{Fe^{2+}}\right)^2 \left(a_{O_2}\right)^{1/2}} \tag{1-35}$$

Solute Activity and Activity Coefficients

The equilibrium constant for a reaction is formulated in terms of the activities (effective concentrations) of the dissolved constituents and not their total dissolved concentrations. This allows the equilibrium constant to be independent of ion shielding effects between dissolved constituents. Ion shielding is a model of solution interactions that attributes the decrease in effective concentration of dissolved species as the ionic strength of the solution increases to the shielding effect of the ions on each other. The negatively charged anions in solution are attracted to the positively charged cations and vice versa, forming a cloud of oppositely charged particles around each ion. This structure in the solution somewhat inhibits chemical reactions from taking place, and the amount of inhibition is a function of the overall ionic strength (sum of the cations and anions) of the solution. The ion shielding model of solution interactions explains such observations as the increase of calcite solubility as a result of adding a salt such as NaCl to a solution in contact with calcite. The salt dissolves releasing sodium (Na^+) and chloride (Cl^-) to solution, thereby providing cations and anions to shield dissolved calcium and carbonate from each other. This shielding reduces the effective concentrations of calcium and carbonate in solution requiring more calcite to dissolve to reach equilibrium with its dissolved constituents (see Section 3.2 for example calculations).

The relationship between the activity of a dissolved constituent and its total dissolved concentration is provided by the following:

$$a_i = \gamma_i c_i \tag{1-36}$$

where a_i is the activity of component i, γ_i is the activity coefficient of component i, and c_i is the total dissolved concentration (moles/L) of component i.

Because of ion shielding the activity of a dissolved species is usually less than its actual dissolved concentration; therefore the activity coefficient (γ) is less than one. Ion shielding is a function of the concentration and valence of the dissolved ions, which is quantified in the ionic strength (I) of a solution:

$$I = 1/2 \, \Sigma c_i \left(z_i\right)^2 \tag{1-37}$$

where c_i is the total dissolved concentration (moles/L) of component i, and z_i is the valence of component i.

There are various methods for calculating the activity coefficient for a dissolved component using the ionic strength of the solution. These include the Davies Equation,

Debye-Huckel Equation, and the extended Debye-Huckel Equation.[13] The variety of expressions have been developed to extend the usefulness of the ion shielding approach to quantifying solution interactions to higher ionic strengths. In general, the ion shielding approach is applicable to solutions with an ionic strength of 0.5 or less. A Ca–HCO_3 groundwater with a TDS of 1200 mg/L has an ionic strength of about 0.02, whereas seawater with a TDS of about 35,000 mg/L has an ionic strength of 0.7. For most groundwater situations, the ion shielding approach is adequate for understanding the geochemistry of the system; however, in some groundwater situations (seawater intrusion zones and near salt deposits) the ion interaction theory[14] of solution processes is required.

Hem[2] provides a figure relating activity coefficients for typical groundwater constituents to ionic strengths. The values were calculated using the Debye-Huckel equation at a temperature of 25°C. At an ionic strength of 0.01, the activity coefficients for the monovalent species are near 0.91, for divalent species 0.67, and for trivalent species 0.45. These correspond to effective concentrations of 91, 67, and 45% of the total dissolved concentration of these species. Ion shielding can have a major effect on the solubility of minerals containing these elements, as well as other reactions occurring in the system.

A second correction that must be taken into account when calculating solution processes and mineral equilibrium is aqueous complexation. The reactions on which the equilibrium constants are based are in terms of the activities of the "free" uncomplexed ions. In the case of calcite, we wrote the reaction (Eq. 1-26) in terms of the mineral and dissolved calcium and carbonate. In solution, carbonate is present as free, uncomplexed carbonate (CO_3^{2-}) and carbonate complexed with hydrogen to form bicarbonate (HCO_3^-) and carbonic acid (H_2CO_3). The total concentration of these three species is referred to as the total inorganic carbon concentration. When we evaluate whether or not a solution is at equilibrium with calcite and use the equilibrium constant given in Equation 1-28, we can only use the free carbonate concentration (adjusted by the activity coefficient) and not the total inorganic carbon concentration. Ignoring aqueous complexation can have a major effect on our calculation. For example, if we measure an alkalinity of 200 mg/L as $CaCO_3$ for a water sample at a pH of 8, the combined carbonate/bicarbonate concentration is about 122 mg/L as HCO_3^-, but the carbonate component of inorganic carbon is only 0.6 mg/L.

The formation of other solution complexes further lowers the concentration of free species. If sulfate is present in the solution that we are equilibrating with calcite, some of the dissolved calcium will form a $CaSO_4^0$ dissolved species, which will require more calcite to dissolve to reach equilibrium and result in a higher total dissolved calcium concentration in solution compared to a system in which sulfate or other complexants are absent. Chapter 2 (Section 2.1) provides additional details on the effects of aqueous complexation on the activities of dissolved constituents, and Section 5.2 describes the use of computer codes to calculate the concentrations and activities of the aqueous species.

Temperature Effects

The equilibrium constants that have been determined for the reactions of interest to geochemists are normally tabulated for systems at 25°C. These constants must be recalculated using the temperature of the aquifer or other system of interest. This requires an accurate aquifer temperature and a means of converting the equilibrium constant to a value for that temperature. The van't Hoff equation is normally used to correct equilibrium constants for temperature:

$$\log_{10} \frac{K_2}{K_1} = \frac{\Delta H_R^0}{2.303R} \left(\frac{T_2 - T_1}{T_2 T_1} \right)$$

(1-38)

where K_2 = equilibrium constant at temperature of interest, T_2 (degrees Kelvin), K_1 = equilibrium constant at 25°C (T_1, 298K), ΔH_R^0 = change in the enthalpy of the reaction (calories/mole), and R = gas constant = 1.987 calories/degree/mole.

1.3. SOIL/ROCK COMPOSITION

The major source of dissolved constituents to groundwater is the solid material contacted by the water. All solids dissolve to some extent into groundwater, thereby mobilizing their constituents. Some solids may form in the groundwater environment or serve as adsorbing substrates thereby acting as sinks for dissolved constituents. The solids that remove dissolved constituents from water will affect the mobility of those constituents and may limit the amount in solution to an equilibrium concentration value. The various types of solid material in aquifer systems and their importance to groundwater composition and chemistry are described in this section.

Rocks and Primary Minerals

The three basic rock types are classified by geologists as igneous, metamorphic, and sedimentary. Igneous rocks are formed from the solidification of molten material (magma) either deep within the earth's crust to form granitic/gabbroic (silica rich/silica poor) rock types or at the surface to form basaltic/andesitic rock types. Metamorphic rocks (e.g., gneiss and schist) form at relatively high temperatures and pressures compared with earth surface conditions by the recrystallization of a rock without its undergoing melting. Sedimentary rocks form under surface or near-surface conditions. They are derived either from the accumulation of the physical weathering products of other rock types producing, for example, sandstones and shales, or from the accumulation of the shells and tests of organisms resulting in rocks such as limestone and chert. Approximately 75% of the continental rocks near the earth's surface are sedimentary rocks or metamorphic rocks derived from sedimentary material with the remaining 25% being igneous rocks.[15] Aquifers can be present in all the different rock types. In general, porous, intergranular flow will predominate in poorly cemented sedimentary rocks, and fracture flow will predominate in igneous and metamorphic rocks.

Rocks are composed of minerals and amorphous solids. A mineral is a crytalline solid with a uniform composition and a regularly repeating structure. For example, the important silicate (Al–Si) group of minerals is composed of various combinations of silica tetrahedra in linear forms (hornblendes and pyroxenes), sheets (biotite and muscovite), and individual tetrahedra (olivines). The principle rock-forming minerals for the three basic rock types are given in Table 1-3. Non-crystalline solids may form a large component of some rocks such as glass in basalt and amorphous silica in chert.

TABLE 1-3.

COMMON ROCK-FORMING MINERALS		
Igneous rocks	**Metamorphic rocks**	**Sedimentary rocks**
Quartz	Quartz	Clays
Feldspars	Calcite	Calcite
Pyroxene	Clays	Quartz
Amphibole	Hematite	Ferrihydrite
Biotite	Source rock minerals	Dolomite
Muscovite		Source rock minerals
Olivine		

Minerals and amorphous solids are composed of chemical elements. Although almost all of the elements on the Periodic Table, other than some of the gases, are found in one or more solids found in the earth's crust, the major components of the earth's crust are limited to a relatively few elements. These elements and their concentration in the near surface environment are listed in Table 1-4. Note that some of the major constituents of

TABLE 1-4.

TYPICAL ROCK AND SOIL CONCENTRATION RANGES

Constituent	Soil/rock (ppm)	Average soil (ppm)
Major Cations		
Calcium	7,000–500,000	13,700
Magnesium	600–6,000	5,000
Sodium	750–7,500	6,300
Potassium	400–30,000	8,300
Silicon	230,000–350,000	320,000
Iron	7,000–550,000	38,000
Aluminum	10,000–300,000	71,000
Major Anions		
Chloride	20–900	100
Sulfate	30–10,000	700
Inorganic Carbon	100–50,000	20,000
Nitrate	200–4,000	1,400

From Lindsay, W.L. *Chemical Equilibria in Soil*, John Wiley & Sons, New York, 1979, and Appello, C. A. J. and Postma, D, *Geochemistry, Groundwater, and Pollution*, A. A. Balkoma, Rotterdam, 1994.

the rocks are also major constituents in groundwater (e.g., calcium, magnesium, sodium, magnesium, and inorganic carbon); however, the rock components of highest concentration (silicon, aluminum, and iron) are not typically major dissolved constituents in groundwater. This occurrence is primarily due to the low solubility of secondary minerals of silicon, aluminum, and iron. This will be discussed further in Section 1.4.

Chemical Weathering
Complete dissolution of the minerals of a rock or the partial alteration of the composition of a mineral due to leaching of its components is termed chemical weathering. These reactions occur because the initially dilute rainwater or snowmelt that contacts the rock is not in chemical equilibrium with the minerals and amorphous solids comprising the rock. The water/rock system reacts and moves toward equilibrium by dissolving or leaching constituents out of the rock into the solution. The primary minerals in the rock may have formed under pressure, temperature, and water vapor conditions far different from that present near the earth's surface. These minerals (especially the olivine, pyroxene, and hornblende classes) will never be in equilibrium with groundwater at the pressure and temperature regimes of normal aquifers. These minerals will dissolve irreversibly, releasing their constituents into the water. Depending on site-specific conditions, complete dissolution of a mineral may take hundreds to millions of years. Primary minerals in some rock types (e.g., calcite in limestone and gypsum in a salt deposit) can and do form under aquifer conditions; therefore their dissolution reactions are reversible, and they may equilibrate with the groundwater.

The alteration products of weathering are dissolved constituents and secondary minerals that can form in the groundwater environment. These are the minerals that are most important to the study of groundwater geochemistry because they are reactive in the aquifer environment. They can limit solution concentration for their constituents and also tend to provide the most common substrates for adsorption reactions. The more common classes of reactive minerals and representative minerals within each class are listed in Table 1-5. The term reactive mineral is used to denote a mineral that will dissolve into or precipitate from groundwater in a reasonable period of residence time for the water in the aquifer. Although reaction rates for most minerals are not known, and for those that have been studied the rate is often a function of many variables, the minerals listed in Table 1-5 have been found to occur in aquifers and to be in thermodynamic equilibrium with the composition of the groundwater. Appendix A contains a more complete list of minerals that might limit solution concentrations for various elements.

TABLE 1-5.

COMMON REACTIVE MINERALS IN SOIL AND AQUIFER ENVIRONMENTS

Carbonates
Calcite [$CaCO_3$]
Dolomite [$CaMg(CO_3)_2$]
Siderite [$FeCO_3$]
Rhodochrosite [$MnCO_3$]
Magnesian Calcite [$(Ca,Mg)CO_3$]

Oxides/hydroxides
Ferrihydrite [$Fe(OH)_3$]
Goethite [$FeOOH$]
Gibbsite [$Al(OH)_3$]
Nsutite [MnO_2]

Sulfates
Gypsum [$CaSO_4 \cdot 2H_2O$]
Alunite [$KAl_3(SO_4)_2(OH)_6$]
Jarosite [$KFe_3(SO_4)_2(OH)_6$]
Jurbanite [$AlSO_4(OH)_6$]

Sulfides
Pyrite [FeS_2]
Mackinawite [FeS]
Orpiment [As_2S_3]

Silica
Chalcedony
Amorphous silica

Silicates
Clays
Zeolites

The clay minerals and metal oxides are particularly important because they not only form in the aquifer environment, but they provide very reactive surfaces on which dissolved constituents can be adsorbed from water. The minerals, amorphous oxides, and organic matter present in an aquifer that participate in adsorption/desorption processes each have separate affinities for the dissolved constituents, and, based on the amount of the substrate, they each have different capacities for adsorption. The process of adsorption/desorption is described in detail in Section 3.1.

Soil Zone

The skin of the earth that supports plant life is one of the most important components in the study of groundwater chemistry. It is in the soil zone that water first enters the subsurface and comes into close contact with inorganic and organic solids. The inorganic material may be fairly fresh rock material as found on the steeper slopes of a mountain, or, more likely, it is weathered material composed of fairly reactive secondary minerals that may equilibrate with the infiltrating water.

The natural organic matter in the soil is composed of a continuum from readily degradable plant debris and roots of dead vegetation to humic substances that have been partially decomposed by soil processes. The humic substances have been classified as humic and fulvic acids with an average worldwide composition in soil of $C_{187}H_{186}O_{89}N_9S$ (humic acid) and $C_{135}H_{182}O_{95}N_5S_2$ (fulvic acid).[16] Solid organic matter plays an important role in the movement of dissolved organic matter because of the attraction of the solid phase for the dissolved phase and consequent removal of dissolved organic carbon from solution by adsorption. The solid organic matter also adsorbs metals from the water, and where inorganic adsorbents are present in low concentration compared with organic adsorbents, the solid organic carbon may be an important control on the mobility of trace metals.

Organic carbon in the soil also affects the inorganic chemistry of groundwater because the oxidation of the organic matter produces carbon dioxide gas. Carbon dioxide gas reacts with the water to produce carbonic acid and the other inorganic carbon species bicarbonate and carbonate. The production of carbonic acid makes the water a more aggressive weathering solution and the increase in the level of anionic carbon species enhances ion complexation. This process is discussed in greater detail in Chapter 2.

Figure 1-8 shows in schematic form the types of chemical reactions that are particularly important in the soil zone. The combination of the presence of reactive minerals, relatively high levels of organic matter, and an aggressive CO_2-rich solution makes the soil zone a major contributor to the composition of groundwater. This is true even though the thickness of the soil is a relatively minor component of the overall groundwater flow path.

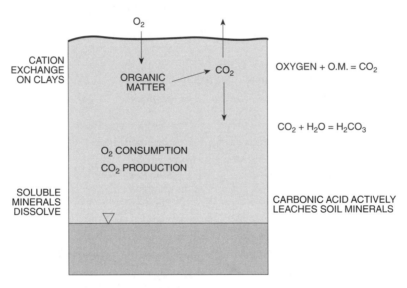

Figure 1-8   Processes in the soil and vadose zones.

1.4. WATER/ROCK PROCESSES CONTROLLING SOLUTION COMPOSITION

The composition of groundwater is a function of the sources and sinks of chemical elements along the groundwater flow path. The relative concentrations of dissolved constituents are determined by the available supply from the solid phases and the solubility of secondary minerals formed from weathering processes. If a chemical element is present only at small concentrations in the aquifer solids, then groundwater will have little opportunity to accumulate the element in solution unless the flow path and groundwater residence times are long. Chlorine is an example of an element that is not generally present in high concentrations in aquifer solids, but which might be found in relatively high concentrations in deeper aquifers where residence times may be on the order of thousands of years. Some of the other halide elements such as bromine and iodine may also accumulate in groundwater, but they are generally present in solids at even lower concentrations than chlorine; therefore they usually remain as minor constituents in solution. The situation for halides illustrates the case where solution concentration is not controlled by an aquifer solid. Concentration in groundwater of the halide ions is typically low because the supply is low; however, given sufficient time, these elements may accumulate in groundwater.

If a chemical element is present in high concentrations in aquifer solids, then it has the potential for occurring as a major component of the groundwater. In an aquifer the primary rock-forming minerals will be leached and dissolved by the groundwater to produce more stable secondary minerals and release some components to solution. For example, depending on local conditions, the weathering of biotite might form the secondary minerals kaolinite [$Al_2Si_2O_5(OH)_4$] and ferrihydrite [$Fe(OH)_3$]. Potassium, magnesium, and silica may also be released into solution according to the following reaction:

$$K_2\left[Si_6Al_2\right]Mg_4Fe_2O_{20}(OH)_4(biotite) + 10H^+ + 6H_2O + 1/2O_2 \rightarrow$$
$$Al_2Si_2O_5(OH)_4 + Fe(OH)_3 + 4H_4SiO_4^0 + 4Mg^{2+} + 2K^+ \qquad (1\text{-}39)$$

Ferrihydrite and kaolinite are secondary minerals that will equilibrate with the groundwater containing their dissolved components iron, aluminum, and silicon per the following equilibrium reactions:

$$Fe(OH)_3 + 3H^+ \leftrightarrow Fe^{3+} + 3H_2O \qquad (1\text{-}40)$$

$$Al_2Si_2O_5(OH)_4 + 5H_2O \leftrightarrow 2Al(OH)_3^0 + 2H_4SiO_4^0 \qquad (1\text{-}41)$$

The solubility of biotite (i.e., the amount that will dissolve in solution) does not control the concentrations of its products because biotite is not a stable mineral under aquifer conditions. Biotite dissolves irreversibly, and the amount of its products (both solid phases and elements released to solution) in the system is mainly a function of the rate at which biotite weathers and the amount of time weathering has occurred. The secondary minerals that form from the alteration of the primary minerals can limit the solution concentration of their constituents. In this case ferrihydrite will control the dissolved iron concentration and kaolinite may limit the solution concentration of one of its components. Each mineral can only limit the solution concentration for one of its components. In the example of biotite weathering, the aluminosilicate clay mineral kaolinite is formed and excess silica ($H_4SiO_4^0$) is shown dissolved in the groundwater (Eq. 1-39). If the silica concentration in solution is not limited by some other mineral, then the formation of kaolinite will limit the amount of aluminum in solution because the other component of kaolinite, silica, is present in solution in excess of the amount of aluminum. As biotite dissolves, the kaolinite reaction (Eq. 1-41) will be driven toward the left and kaolinite will precipitate removing both aluminum and silica from solution, but overall the silica solution concentration will increase as the aluminum concentration is depressed. This example of mineral control on solution concentrations illustrates the Phase Rule, which states that each mineral phase can only control the dissolved concentration of one of its components in solution.

Barite can also be used to illustrate the importance of the Phase Rule on solid phase control of solution concentration. The barite dissolution reaction can be written as follows:

$$BaSO_4 \leftrightarrow Ba^{2+} + SO_4^{2-} \qquad K = 10^{-9.98} \qquad (1\text{-}42)$$

If the sulfate concentration in the groundwater is 10^{-5} moles/L and the groundwater is in equilibrium with barite, then, ignoring the correction for activity, the dissolved barium concentration will be $10^{-4.98}$ moles/L (=$10^{-9.98}/10^{-5}$). If the sulfate concentration were fixed at 10^{-4}, then equilibrium with barite would limit the barium concentration to $10^{-5.98}$ moles/L. Conversely, if the barium concentration was varied independently, the sulfate concentration in the water would be fixed by equilibrium with barite. However, the single solid phase barite cannot fix the dissolved concentration of both of its components in solution unless it is the only source of barium and sulfate in the system. This is unlikely in natural waters.

The dissolved concentrations of the major elements in silicate rocks (silicon, aluminum, and iron) are normally limited in groundwater by the formation of reactive minerals of low solubility. These solids are amorphous or partially crystalline silica (SiO_2), amorphous $Al(OH)_3$ or gibbsite, clay (aluminosilicate) minerals, and ferrihydrite [$Fe(OH)_3$]. Because of the low solubility of these solids under normal groundwater pH and Eh conditions, the major elements in the aquifer solids are not major ions in the groundwater solution. The major cations (Ca^{2+}, Mg^{2+}, Na^+, and K^+) and anions (HCO_3^-/CO_3^{2-}, SO_4^{2-}, Cl^-) in solution, except for chloride, are those present in moderate concentrations in typical host rocks and that form relatively soluble secondary minerals (carbonates and sulfates). Mineral solubility and other water/rock interactions that affect solution composition are discussed further in Chapter 3.

REFERENCES

1. Masters, R.W., National Ground Water Information Center, 1996.
2. Hem, J.D., *Study and Interpretation of the Chemical Characteristics of Natural Water,* U.S. Geological Survey, 1989.
3. Fritz, S.J., A survey of charge-balance errors on published analyses of potable ground and surface waters, *Ground Water,* 32, 539–546, 1994.
4. Kennedy, V.C., Zellweger, G.W., and Jones, B.F., Filter pore-size effects on the analysis of Al, Fe, Mn and Ti in water, *Water Resour. Res.,* 10, 785–790, 1974.
5. Piper, A.M., A graphic procedure in the geochemical interpretation of water analyses, *AGU Trans.,* 25, 914–923, 1944.
6. White, D.E., Hem, J.D., and Waring, G.A., Chemical composition of subsurface water, in *Data of Geochemistry,* 6th ed., U.S. Geological Survey, 1963.
7. A.P.H. Association, A.W.W. Association, and W.P.C. Federation, *Standard Methods for the Examination of Water and Wastewater,* American Public Health Association, Washington, DC, 1989.
8. Litaor, M.I. and Thurman, E.M., Acid neutralizing processes in an alpine watershed Front Range, Colorado, U.S.A. I. Buffering capacity of dissolved organic carbon in soil solutions, *Appl. Geochem.,* 3, 645–652, 1988.
9. Kehew, A.E. and Passero, R.N., pH and redox buffering mechanisms in a glacial drift aquifer contaminated by landfill leachate, *Ground Water,* 28, 728–737, 1990.
10. Lindberg, R.D. and Runnells, D.D., Ground water redox reactions: an analysis of equilibrium state applied to Eh measurements and geochemical modelling, *Science,* 225, 925–927, 1984.
11. Barcelona, M.J. and Holm, T.R., Oxidation-reduction capacities of aquifer solids, *Environ. Sci. Technol.,* 25, 1565–1572, 1991.
12. Nikolaidis, N.P. et al., Vertical distribution and partitioning of chromium in a glaciofluvial aquifer, *GWMR,* 150–159, 1994.
13. Stumm, W. and Morgan, J.J., *Aquatic Chemistry,* Wiley-Interscience, New York, 1970, pp. 1–583.
14. Pitzer, K.S., Thermodynamics of electrolytes. I. Theoretical basis and general equations, *J. Phys. Chem.,* 77, 268–277, 1973.
15. Stokes, W.L. and Judson, S., *Introduction to Geology,* Prentice-Hall, Englewood Cliffs, NJ, 1968, pp. 1–530.
16. Sposito, G., *The Chemistry of Soils,* Oxford University Press, New York, 1989.

2 SOLUTION, REDOX, AND GAS EXCHANGE PROCESSES

An understanding of the solution component of natural systems requires information about the chemical processes that occur in this phase of the system. Aqueous complexation reactions, which partition the dissolved elements among the species in solution and can have a major impact on effective concentration, are described in this chapter. Redox reactions that shift the relative concentrations of redox-sensitive species in solution are also described. Both aqueous complexation and redox reactions may be affected by gasses dissolving into or exsolving from solution. Section 2.3 covers the gas exchange process and its impact on solution concentration.

2.1. AQUEOUS COMPLEXATION

In the discussion of solute activity (Section 1.2) it was mentioned that aqueous complexation reactions must be taken into account when comparing dissolved concentrations with expected solution activities at thermodynamic equilibrium. These comparisons are made to help identify controls on solution composition. Because the equilibrium constants have been developed in terms of the activities of free, uncomplexed ions, the measured analytical concentrations in groundwater must be converted to activities of these specific ions to allow for appropriate comparisons.

For example, most of the equilibrium constants for the carbonate minerals are written in terms of the major cation in the mineral and carbonate, (calcite: $a_{Ca^{2+}}$ and $a_{CO_3^{2-}}$; siderite: $a_{Fe^{2+}}$ and $a_{CO_3^{2-}}$; witherite: $a_{Pb^{2+}}$ and $a_{CO_3^{2-}}$; etc.). If the alkalinity of the water sample is used to derive the carbonate concentration, it must first be corrected for noncarbonate alkalinity as discussed in Section 1-1 and then that portion of the inorganic carbon alkalinity attributable to bicarbonate must be subtracted to arrive at the value of carbonate in solution. Alternatively, if the total inorganic carbon (TIC) concentration of the water is known, then the concentrations of carbonic acid and bicarbonate must be subtracted from the TIC concentration to calculate the carbonate concentration (Example 2-1). However, in each case, the value calculated for carbonate from the alkalinity measurement or TIC is the total dissolved carbonate concentration in solution and not just the free, uncomplexed dissolved carbonate concentration, which is required for comparison with the mineral equilibrium constants. To calculate the free, uncomplexed carbonate concentration, the concentration of dissolved carbonate complexes must be determined using the association constants for these complexes. In typical groundwater, carbonate forms dissolved complexes with calcium and magnesium; therefore the concentration of these complexes must be calculated and subtracted from the total carbonate concentration to determine the free, uncomplexed carbonate in solution. The free, uncomplexed carbonate concentration can then be multiplied by its activity coefficient to determine the activity of carbonate. This number is used with the activity of the major cation in a mineral to compare with the equilibrium constant of that mineral to determine if the solution is in equilibrium with the mineral (Section 3.2).

Example 2-1. Calculation of carbonate activity

Total inorganic carbon (TIC) concentration = Total carbonate concentration $(CO_{3,T}^{2-})$ +
    Total bicarbonate concentration $(HCO_{3,T}^-)$ + Total carbonic acid concentration $(H_2CO_{3,T})$

Total carbonate concentration $(CO_{3,T}^{2-})$ = TIC concentration –
    Total bicarbonate concentration – Total carbonic acid concentration

$$CO_{3,T}^{2-} = C_{CO_3^{2-}} + C_{CaCO_3^0} + C_{MgCO_3^0} + \ldots$$

$$C_{CO_3^{2-}} = CO_{3,T}^{2-} - C_{CaCO_3^0} - C_{MgCO_3^0} + \ldots$$

$$a_{CO_3^{2-}} = \left(C_{CO_3^{2-}}\right)\left(\gamma_{CO_3^{2-}}\right)$$

For a water sample with a pH = 8.5, TIC = 120 mg/L, Ca^{2+} = 20 mg/L, and Mg^{2+} = 12 mg/L, the following concentrations are present in solution:

Constituent	Concentration (moles/kg)
TIC	2.0×10^{-3}
$H_2CO_{3,T}$	1.3×10^{-5}
$HCO_{3,T}^-$	1.9×10^{-3}
$CO_{3,T}^{2-}$	5.8×10^{-5}
$C_{CaCO_3^0}$	1.4×10^{-5}
$C_{MgCO_3^0}$	9.5×10^{-6}
$C_{CO_3^{2-}}$	3.4×10^{-5}
$a_{CO_3^{2-}} (\gamma_{CO_3^{2-}} = 0.79)$	2.7×10^{-5}

These adjustments to the carbonate concentration can have a large impact on the value used to compare with the equilibrium constant. Example 2-1 shows the steps required in the calculation and the results for a water sample with a TIC concentration of 120 mg/L as CO_3^{2-}, a pH of 8.5, calcium concentration of 20 mg/L, and a magnesium concentration of 12 mg/L. Of the TIC concentration of 120 mg/L only 3.46 mg/L is present as total carbonate $(CO_{3,T}^{2-})$ and only 2.03 mg/L is present as free, uncomplexed carbonate $(C_{CO_3^{2-}})$. For this solution, the activity coefficient of carbonate is 0.79; therefore the activity of carbonate is 1.6 mg/L. This is only 1.3% of the TIC concentration.

Table 2-1 shows the results of speciation calculations for a typical Ca–HCO_3 groundwater. As expected at a pH of 8.4, most of the inorganic carbon is present as bicarbonate. Other inorganic carbon species present at 1 to 2% of the total inorganic carbon concentration are free, uncomplexed carbonate (CO_3^{2-}); the calcium carbonate complex $(CaCO_3^0)$; and the calcium bicarbonate complex $(CaHCO_3^+)$. The divalent calcium and magnesium cations are complexed appreciably with carbonate, bicarbonate, and sulfate, while the univalent cations (Na^+ and K^+) and anions (Cl^- and NO_3^-) form little or no complexes. The tendency to form dissolved complexes is a function of the electrical charge on the ions, which attracts oppositely charged species. The higher the charge, the greater the likelihood of forming strong solution complexes. Ferric iron (Fe^{3+}) is a case in point because of its strong tendency

TABLE 2-1.

ION SPECIATION AND COMPLEXATION IN TYPICAL GROUNDWATER

Constituent	Concentration (a) (mg/L)	Solution species/ complex	Concentration (b) (%)
Calcium	92	Ca^{2+}	87.5
		$CaSO_4^0$	5.1
		$CaCO_3^0$	4.4
		$CaHCO_3^+$	3.0
Magnesium	34	Mg^{2+}	89.0
		$MgSO_4^0$	4.6
		$MgHCO_3^+$	3.5
		$MgCO_3^0$	3.0
Sodium	8.2	Na^+	99.4
		$NaHCO_3^0$	0.2
		$NaSO_4^-$	0.2
Potassium	1.4	K^+	99.7
		KSO_4^-	0.3
Total Inorganic	330	HCO_3^-	93.0
Carbon		$CaCO_3^0$	1.8
		CO_3^{2-}	1.5
		$CaHCO_3^+$	1.3
Chloride	9.6	Cl^-	100
Nitrate	13	NO_3^-	100.0
Iron	0.09	$Fe(OH)_4^-$	67.1
		$Fe(OH)_3^0$	23.9
		$Fe(OH)_2^+$	9.1
		Fe^{3+}	<<1

[a] Data from Snoeyink and Jenkins, 1980.
[b] Calculated using MINTEQA2 (Allison et al. 1991)

to form hydroxylated species. As shown in Table 2-1, at a pH of 8.4 the dominant dissolved iron species in solution are $Fe(OH)_4^-$, $Fe(OH)_3^0$, and $Fe(OH)_2^+$. The free, uncomplexed ferric ion is present in solution at only $1.6 \times 10^{-10}\%$ of the total dissolved iron concentration. Other typical trace metals in groundwater such as aluminum, silicon, and chromium (III) also form strong hydroxyl complexes in water at normal groundwater pH values. It is especially important to take aqueous complexation into account for these elements when evaluating metal mobility and possible controls on solution concentration.

Because of the large number of components typically present in groundwater, the number of species that are formed from these components, and the interdependence of the calculations, it is not generally practical to make these complexation calculations by hand. Computer codes such as MINTEQA2 and PHREEQE are available to perform these calculations provided that accurate concentration, temperature, and pH data, at a minimum, are available. The use of computer codes and their availability are discussed in Chapter 5.

2.2. OXIDATION/REDUCTION PROCESSES

Oxidation/reduction processes involve the transfer of electrons from one element to another, thereby changing the valence of the elements. These processes are important in the study of groundwater geochemistry because many of the elements in the system are redox-sensitive and are susceptible to electron transfer. The solubilities of minerals formed from these elements are dependent on the redox potential, Eh, of the system, which therefore directly affects the mobility of these elements. Furthermore, the toxicity of certain elements is dependent on the redox state. For example, Cr(VI) is more toxic than Cr(III), and As(III) is more toxic than As(V). This discussion of redox processes supplements the text on redox potential (Section 1.1) and field methods for accurate measurement of Eh (Section 9.3).

Oxidation/reduction processes in the subsurface are driven by disequilibrium conditions as are the other geochemical reactions. For example, dissolved oxygen entering the groundwater environment produces a relatively oxidizing solution compared with all of the redox-sensitive elements found in the natural environment. Carbon in organic matter is commonly in the zero or lower valence state [C(0) to C(-IV)], which is not the stable oxidation state of carbon in equilibrium with measurable dissolved molecular oxygen [O_2^0(aq)]. The stable oxidation state of carbon in the presence of molecular oxygen is C(IV), as in CO_2. The disequilibrium between O_2 and organic carbon results in the oxidation of carbon and reduction of oxygen according to the following general reaction:

$$CH_2O + O_2 \leftrightarrow CO_2 + H_2O$$

The half-cell reactions display the transfer of electrons for this oxidation-reduction reaction:

$$CH_2O + 2H_2O \rightarrow CO_2 + 4e^- + 4H^+ \quad \text{oxidation of } C(0) \text{ to } C(IV)$$

$$O_2 + 4e^- + 4H^+ \rightarrow 2H_2O \quad \text{reduction of } O(0) \text{ to } O(-II)$$

In this reaction, molecular oxygen is consumed in the oxidation of the organic matter to carbon dioxide and water. The reaction will continue until equilibrium conditions are established by decreasing to a low concentration either the organic matter or dissolved oxygen in the system. Similar conditions are produced by weathering of primary minerals and the release of reduced metal species such as Fe^{2+} and Mn^{2+} to solution. The metals are oxidized to a higher valence state by dissolved oxygen until equilibrium is achieved between the oxidizing groundwater and the dissolved redox-sensitive species.

Iron is one of the most important redox-sensitive elements in the environment and can be used to illustrate oxidation-reduction processes. An exchange of electrons occurs when solid, elemental iron (Fe^0(s)) is oxidized by molecular oxygen (O_2) to the ferrous iron [Fe(II)] redox state according to the following reaction:

$$Fe^0(s) + 1/2 O_2 + H^+ \rightarrow Fe^{2+} + H_2O \qquad (2\text{-}1)$$

For this reaction to occur, the solid elemental iron must be contacted by water containing dissolved oxygen. The reaction converts some of the dissolved oxygen and hydrogen ion to water and produces dissolved ferrous iron. The transfer of electrons in this reaction can be seen from the half-cell reactions, which show the oxidation (loss of electrons and increase in valence) of iron and the reduction (gain of electrons and decrease in valence) of oxygen:

$$Fe^0(s) \rightarrow Fe^{2+} + 2e^- \quad \left(\text{oxidation of } Fe^0(s) \text{ to } Fe^{2+}\right) \qquad (2\text{-}2)$$

$$1/2 O_2^0 + 2H^+ + 2e^- \rightarrow H_2O \quad \left[\text{reduction of } O(0) \text{ to } O(-II)\right] \qquad (2\text{-}3)$$

Note that adding together the two half-cell reactions eliminates the two electrons and produces the complete oxidation-reduction reaction (Eq. 2-1). Redox reactions are coupled reactions that must include an oxidation with a reduction reaction. Free electrons are not produced by the coupled redox reactions, but are simply transferred between species. There are no free electrons in the environment.

In water, the dissolved ferrous iron produced by oxidation of elemental iron may be further oxidized by dissolved oxygen to the ferric iron [Fe(III)] redox state, which may then precipitate as a ferric hydroxide mineral according to the following:

$$Fe^{2+} + 1/4\,O_2 + 5/2\,H_2O \rightarrow Fe(OH)_3(s) + 2H^+ \tag{2-4}$$

In this case, the valence of iron is changed from ferrous [Fe(II)] to ferric [Fe(III)] iron and only one electron is transferred per atom by this oxidation/reduction process. This can be seen from the half-cell reactions:

$$Fe^{2+} + 3H_2O \rightarrow Fe(OH)_3(s) + 3H^+ + e^- \quad \left(\text{oxidation of } Fe^{2+} \text{ to } Fe^{3+}\right) \tag{2-5}$$

$$1/4\,O_2^0 + H^+ + e^- \rightarrow 1/2\,H_2O \quad \left[\text{reduction of } O(0) \text{ to } O(-II)\right] \tag{2-6}$$

A representative reaction for what happens when elemental iron in cast iron or steel is in contact with water can be derived by combining reactions 2-1 and 2-4:

$$Fe^0 + 3/4\,O_2 + 3/2\,H_2O \rightarrow Fe(OH)_3(s) \tag{2-7}$$

The dissolved oxygen in the water oxidizes the reduced iron Fe(0) to Fe(III) in the form of ferrihydrite, commonly described as rust. The formation of ferrihydrite typically limits the dissolved iron concentration in groundwater when dissolved oxygen is present.

The use of roman numerals with the element denotes all of the species of that valence state; for example, Fe(III) includes all of the ferric iron species Fe^{3+}, $Fe(OH)^{2+}$, $Fe(OH)_2^+$, $Fe(OH)_3^0$, etc. dissolved in solution. In the groundwater environment, dissolved species of both Fe(II) and Fe(III) will be present. The relative proportion of ferrous and ferric iron in solution is a function of the redox potential of the system. The redox potential of a system can be thought of as a measure of its relative oxidation/reduction state. A high positive potential will favor the more oxidizing valence states of an element, while lower potentials favor the more reduced valence states. The redox potential is a relative measure because what may be oxidizing for one element might be reducing for another. For example, the reduction of oxygen from molecular oxygen $[O_2^0, O(0)]$ to O^{2-} $[O(-II)]$ occurs at a very high redox potential ($\approx$800 mv at pH 8), whereas the reduction of sulfur in sulfate (S(VI) to sulfide (S(-II) occurs at a low redox potential ($\approx$–300 mV at pH = 8). At a pH of 8 and a redox potential of 400 mV, we would expect to find very little of the more oxidized molecular oxygen in solution, but we would expect to find most of the dissolved sulfur in its oxidized sulfate form. (Note that oxidation/reduction processes do not necessarily involve oxygen. All that is required is a transfer of electrons between elements. Oxygen happens to be one of the strongest and most common oxidizing agents in the atmosphere and shallow waters, hence its use as part of the name for the process).

The half-cell reactions can be used to calculate the relative proportion of redox-sensitive species of an element in solution at a given redox potential. For example, the half-cell reaction between ferrous and ferric iron is as follows:

$$Fe^{2+} \leftrightarrow Fe^{3+} + e^- \quad E^0 = +0.77 \text{ volts} \tag{2-8}$$

In this reaction, Fe^{2+} and Fe^{3+} are the free, uncomplexed species of ferrous and ferric iron, respectively. They are not the total dissolved concentrations of Fe(II) and Fe(III). E^0 is the standard potential for the reaction at 25°C with Fe^{2+} and Fe^{3+} present in solution at unit activities (i.e., 1 mole/liter). The standard potential is measured relative to the hydrogen electrode; therefore a measurement of solution Eh will allow us to determine the relative proportions of the dissolved iron species. If the Eh of a solution is 0.77 volts, then the activities of dissolved Fe^{2+} and Fe^{3+} will be equal if the solution is in redox equilibrium. If the activity of the oxidized iron species (Fe^{3+}) is greater than that of the reduced iron species (Fe^{2+}), then the redox potential will be higher than 0.77 volts, and vice versa. We can calculate the Eh for any ratio of the dissolved species using the Nernst Equation:

$$Eh = E^0 + 2.303 \frac{RT}{nF} \log \frac{(\text{activity product of reaction products})}{(\text{activity product of reactant species})} \qquad (2\text{-}9)$$

where, E^0 = standard potential for the reaction written as reduced $\leftrightarrow$ oxidized + ne^-, R = gas constant (1.987 cal/deg mole), T = temperature (K), n = number of electrons in the reaction, and F = Faraday's constant (23,061 cal/volt).

At 25°C, the Nernst Equation reduces to the following:

$$Eh = E^0 + \frac{0.059}{n} \log \frac{(\text{activity product of reaction products})}{(\text{activity product of reactant species})} \qquad (2\text{-}10)$$

In the case of the ferrous/ferric iron redox couple (Eq. 2-8), the reaction product is Fe^{3+}, the reactant species is Fe^{2+}, and E^0 = 0.77 volts; therefore,

$$Eh(V) = 0.77\ V + 0.059 \log \frac{a_{Fe^{3+}}}{a_{Fe^{2+}}} \ (25°C)$$

Table 2-2 provides ratios of the activities of ferrous and ferric iron at different Eh values. As expected, Eh values below 0.77 volts favor the reduced, ferrous form of iron; however, it should be kept in mind that even at the lowest Eh values there is still some of the oxidized, ferric species of iron present as required by the equilibrium condition between the species. Also, the Eh value only provides specific information on the activities of the free, uncomplexed species. Because ferrous and ferric iron form numerous hydrolysis species of different strengths in water, the ratio of the total dissolved ferrous and ferric iron in solution is quite different from the ratios of the activities of the free ions. As shown in Table 2-2, the very strong ferric species $Fe(OH)_2^+$ distorts the total iron ratio to favor Fe(III), and it is not until the Eh is fairly low (about 0.3 volts) that total Fe(II) exceeds total Fe(III).

TABLE 2-2.

EFFECT OF EH ON THE DISTRIBUTION OF IRON SPECIES

Eh(V)	$a_{Fe^{3+}}/a_{Fe^{2+}}$	Fe(III)/Fe(II) (pH=7)	Dominant Species
0.77	1	2.3×10^8	$Fe(OH)_2^+$
0.65	0.01	2.2×10^6	$Fe(OH)_2^+$
0.48	10^{-5}	2.9×10^3	$Fe(OH)_2^+$
0.18	10^{-10}	2.4×10^{-2}	Fe^{2+}
−0.12	10^{-15}	2.0×10^{-7}	Fe^{2+}

pH Dependent Redox Reactions

Many redox reactions are pH dependent because the hydrogen ion is part of the oxidation-reduction reaction. If the oxidation of iron is written in terms of ferrous iron oxidation to ferric iron and precipitation of ferric hydroxide, the half-cell reaction would be as follows:

$$Fe^{2+} + 3H_2O \rightarrow Fe(OH)_3 + 3H^+ + e^- \quad E^0 = +0.98 \text{ volts} \tag{2-11}$$

The Nernst Equation can be used to write this reaction in terms of Eh as follows:

$$Eh = E^0 + 0.059 \log \left(a_{H^+} \right)^3 / \left(a_{Fe^{2+}} \right) \tag{2-12}$$

(Note that the same convention of setting the activity of water and solids equal to one is used with the Nernst Equation as with the equilibrium constant expression (Chapter 1.)

Using the definition of pH as negative log a_H^+, inserting the E^0 value of 0.98 V and setting the activity of the ferrous iron species to 10^{-6} moles/L produces the following:

$$Eh = 0.98 + (0.059)(-3pH) - 0.059 \log(10^{-6})$$

$$= 0.98 + 0.35 - 0.18pH \tag{2-13}$$

$$= 1.33 - 0.18pH$$

This is the equation of a line on an x-y plot of pH versus Eh. Figure 2-1 is a pH-Eh diagram upon which this line is plotted. At any point along the line, the solid ferric hydroxide is in equilibrium with a dissolved activity of Fe^{2+} of 10^{-6} moles/liter. We can also plot lines at which ferric hydroxide is in equilibrium with other activities of Fe^{2+}. The following equations correspond to Fe^{2+} activities of 10^{-4} and 10^{-8} moles per liter:

$$Eh = 1.22 - 0.18pH \quad \left(a_{Fe^{2+}} = 10^{-4} \text{ moles/L} \right) \tag{2-14}$$

$$Eh = 1.45 - 0.18pH \quad \left(a_{Fe^{2+}} = 10^{-8} \text{ moles/L} \right) \tag{2-15}$$

The positions of these lines on Figure 2-1 are above (10^{-8} moles/L) and below (10^{-4} moles/L) the line for an activity of 10^{-6} moles/L. This is consistent with our instinct that at higher Eh values the oxidized form of iron (in this case $Fe(OH)_3s$) would be preferred and it would be less soluble than at lower Eh values. As Eh decreases, at a given pH, the $Fe(OH)_3s$ becomes more soluble and would be in equilibrium with higher dissolved concentrations of ferrous iron. Also shown on the pH-Eh diagram is a line separating the Fe^{2+} and Fe^{3+} ions. In accordance with Equation 2-8, equal activities of these two ions occurs at an Eh of 0.77 volts. The reaction involves only Fe^{2+} and Fe^{3+} and does not include hydrogen, so it is independent of pH and plots as a horizontal line on the diagram.

The upper and lower dashed lines on the diagram are the stability fields for water. They are plots of the following two reactions and Eh equations:

Upper stability limit of water

$$H_2O \leftrightarrow 1/2O_2 + 2H^+ + 2e^- \quad E^0 = 1.23 \text{ volts} \tag{2-16}$$

$$Eh = 1.22 - 0.059pH \quad \left(pO_2 = 0.2 \text{ atm} \right) \tag{2-17}$$

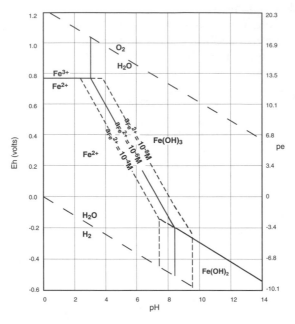

Figure 2-1 Eh-pH diagram for selected iron species.

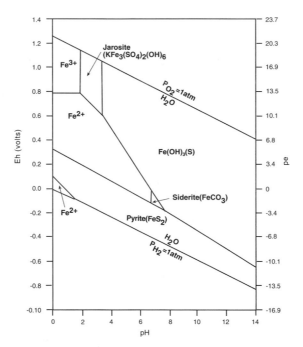

Figure 2-2 pe-ph diagram of the iron system ($a_{Fe} = a_{k+} = 10^{-4}$ M; $a_s = 10^{-2}$ M; $P_{CO_2} = 0.01$ atm).

Lower stability limit of water

$$H_2^0 \leftrightarrow 2H^+ + 2e^- \quad E^0 = 0.00 \text{ volts} \tag{2-18}$$

$$Eh = -0.059pH \quad \left(pH_2 = 1 \text{ atm}\right) \tag{2-19}$$

At Eh values above the upper dashed line, water is not stable and will dissociate to oxygen gas and H^+, while below the lower dashed line hydrogen gas will be produced by the reduction of H^+ in water.

Because of the pH dependence of many of the redox reactions of importance to geochemistry, the use of a combined Eh and pH parameter to represent lines on the pH-Eh diagram would be beneficial. In addition to representing the redox potential as an Eh value, it can also be represented as pe [negative log of electron activity $(-\log a_{e^-})$, moles/L]. Because pH and pe have the same units (moles/L) they can be added together to give a number with consistent units. The pe + pH values for the lines representing the water stability field are derived as follows:

Upper stability limit of water

$$H_2O \leftrightarrow 1/2\,O_2 + 2H^+ + 2e^- \qquad K = 10^{-41.56} \qquad (2\text{--}20)$$

$$\log K = \log \frac{\left(a_{O_2}\right)^{1/2}\left(a_{H^+}\right)^2\left(a_{e^-}\right)^2}{a_{H_2O}} = -41.56$$

$$1/2\log a_{O_2} - 2pH - 2pe = -41.56 \quad \left(a_{H_2O} = 1\right)$$

$$pH + pe = 20.78 - 1/4\log a_{O_2} = 20.73 \quad \left(a_{O_2} = 0.2\right)$$

Lower stability limit of water

$$H_2 \leftrightarrow 2H^+ + 2e^- \qquad K = 1 \qquad (2\text{-}21)$$

$$\log K = \log \frac{\left(a_{H^+}\right)^2\left(a_{e^-}\right)^2}{a_{H_2}} = 0$$

$$-2pII - 2pe - \log\left(a_{H_2}\right) = 0$$

$$pH + pe = 0 \quad \left(a_{H_2} = 1\right)$$

Using a similar approach with the reaction (Eq. 2-11) for the oxidation of ferrous iron to $Fe(OH)_3$ yields a pH/pe equation for this reaction of $3pH + pe = 22.5$. One of the advantages of the pH + pe approach is that field measurements of pH and the redox potential can be added together to provide a redox status for the soil that takes into account the pH dependence of the redox reactions. Eh can be converted to pe by the following equation:

$$Eh\left(\text{millivolts}\right) = \frac{2.303RT}{F}\,pe = 59.2\;pe\;\left(25°C\right) \qquad (2\text{-}22)$$

Eh and pe are often plotted as the y-axis on pH-Eh diagrams as shown on Figure 2-1. The advantages of using the pe for redox equilibrium calculations have been described by Truesdell,[1] and Lindsay[2] has adopted this approach in his book titled *Chemical Equilibria in Soils*.

Although pe is the negative logarithm of the activity of the electron in moles per liter, this is not an absolute value for the activity of the electron. It is the activity of the electron relative to the standard hydrogen electrode, just as Eh is the redox potential relative to the standard hydrogen electrode. The actual electron activity at the lower limit of water stability, where electron activity is the highest, is reported to be about 10^{-80} moles/L,[3] however, this number is not needed for typical geochemical calculations.

Figure 2-2 shows a more complete pe-pH diagram for the iron system that includes potassium, sulfur, and carbonate compounds. On this diagram the boundaries between solids and solutes are at a solute concentration of 10^{-6} moles/L. Diagrams of this type are a convenient method of displaying the stable minerals for particular components of the system over a wide range of pe/pH conditions. For the iron system, the sulfate mineral jarosite is shown to be stable over a fairly small zone of low pH and high Eh. The stability field of ferrihydrite [$Fe(OH)_3$] spans a large portion of the diagram reinforcing the importance of this mineral in natural systems. The iron carbonate mineral siderite has a fairly small field at a partial pressure for CO_2 of 0.01 atm. This field would expand as the P_{CO_2} increases. Under reducing, low Eh conditions, sulfide becomes a greater component of the dissolved sulfur in the water, allowing the iron sulfide mineral pyrite to occupy this portion of the diagram. Under certain conditions of high pH and low Eh, both the Fe(III) mineral ferrihydrite and the Fe(II) mineral pyrite may coexist.

Other important redox-sensitive elements in the groundwater environment are listed in Table 2-3. Many of these elements also form strong hydroxl species in water, and as a consequence, pH must also be considered in calculating dominant dissolved species. Lindsay[2] provides pH-Eh diagrams for many of these elements. As with the calculation of ion speciation and complexation, the complexity and iterative nature of these calculations generally requires the use of a computer code for practical purposes. These computer codes are discussed in detail in Chapter 5.

Unlike aqueous complexation reactions, which are generally considered to reach equilibrium in very short (lab scale) time frames, redox reactions may be very slow. It is commonly found that disequilibrium exists between the concentrations of redox pairs of an element in groundwater and the measured redox potential.[4] As shown above, the equilibrium ratio of redox-sensitive species of an element can be calculated from a measured redox potential using the Nernst Equation. Alternatively, the dissolved concentrations of

TABLE 2-3.

REDOX-SENSITIVE ELEMENTS AND COMMON VALENCE STATES

Element	Redox States
Arsenic	As(III), As(V)
Carbon	C(-IV) to C(IV)
Chromium	Cr(III), Cr(VI)
Copper	Cu(I), Cu(II)
Iron	Fe(II), Fe(III)
Manganese	Mn(II), Mn(III), Mn(IV)
Mercury	Hg(0), Hg(I), Hg(II)
Nitrogen	N(-III), N(0), N(III), N(V)
Oxygen	O(0), O(-II)
Selenium	Se(-II), Se(0), Se(IV), Se(VI)
Sulfur	S(-II) to S(VI)
Uranium	U(IV), U(VI)
Vanadium	V(III), V(V)

the two species that form the redox pair can be measured in a water sample, and the Eh can be calculated using these values in the Nernst Equation. The Eh measured with a platinum (or other inert) electrode can be compared with the Eh value calculated from the activity values to evaluate whether dissolved activities are in equilibrium with the measured Eh. Figure 2-3 shows ranges of measured and computed Eh values for several redox couples of interest in the aquifer environment. If redox equilibrium existed, the values would fall near the dotted line with a slope of one. The large fields of plotted values show the large redox disequilibrium that can exist in natural waters. For example, the measured dissolved oxygen concentrations in groundwaters should give the water a much higher measured Eh value than normally found, and the HS^- and SO_4^{2-} concentrations should produce lower Eh values than generally measured. Whitfield[5] discusses the limitation on the use of the platinum electrode to the Eh region between about 0 to +400 mV at a pH of 7. The surface of the platinum electrode may not respond to the system Eh at redox potentials higher than +400 mV because of the presence of a PtO coating, and it may not be accurate below an Eh of 0 mV because of the formation of a PtS coating. Also, many redox reactions require bacterial mediation to reach equilibrium even in the time frame of groundwater residence in an aquifer. A case in point is the reduction of S(VI) in sulfate to S(-II) in sulfide, which requires not only strongly reducing conditions but the action of sulfur-reducing bacteria to facilitate the reduction process.

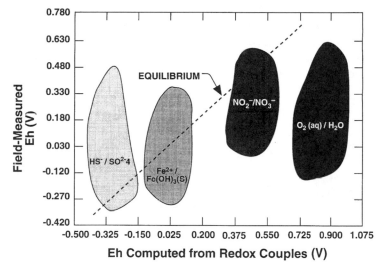

Figure 2-3 $Eh_{measured}$ versus $Eh_{computed}$. (From Lindberg, R.D. and Runnells, D.D., *Science*, 225, 925–927, 1984.)

The fact that redox equilibrium is not common among redox pairs requires that the use of a measured Eh value for quantitative applications to groundwater geochemistry be applied on a selective basis. Just as certain minerals do not equilibrate with groundwater, certain redox pairs do not equilibrate unless necessary conditions are met. Therefore an investigator can only utilize the reactive redox-sensitive species that are appropriate for the groundwater environmental condition. For example, many studies[6-10] have shown the Fe^{2+}/Fe^{3+} and/or $Fe^{2+}/Fe(OH)_3(s)$ redox couples to be in equilibrium with a measured redox potential under a broad range of relatively oxidizing conditions (Eh > 200 mv). It has also been shown that the As(III)/As(V) couple is often in equilibrium with the measured groundwater redox potential.[11] Other redox couples are in equilibrium when conditions allow, such as the NO_3^-/N_2 couple when nitrate-reducing bacteria are present and the SO_4^{2-}/HS^- couple when sulfate reducing bacteria are present.

2.3. SOLUTION/GAS INTERACTIONS

The exchange of constituents between a gas phase and water is an important process affecting the composition of groundwater in the unsaturated zone and at the water table. Surface water that percolates into soil contains gases that have dissolved into the water by contact with atmospheric air. In the soil zone, gases are produced and consumed by interactions with organic matter and minerals. Because of these processes the soil gas vapor phase commonly has a much different composition from atmospheric air, and water in the soil and unsaturated zones will have a different composition and concentration of dissolved gases from surface water. As groundwater moves below the water table and loses contact with a separate vapor phase, it can no longer equilibrate with this phase, but it will initially contain the dissolved gases produced by the last contact of the water with the vapor. Below the water table, the dissolved gas concentration may increase or decrease depending on the types of water/rock interactions that occur in the aquifer. This section discusses the important gases in groundwater and the interactions between phases that impact gas concentrations.

Gas Exchange Processes

Contact of a gas in a vapor phase with water produces a solution containing the gas at some dissolved concentration. Given sufficient time, the concentrations of the gas in the vapor and solution phases will reach equilibrium values. As an example, the interaction of oxygen and carbon dioxide in the atmosphere with water is shown schematically in Figure 2-4, and the reactions can be written as follows:

$$O_2(gas) \leftrightarrow O_2(aq) \; K_{H,O_2} \tag{2-23}$$

$$O_2(gas) \leftrightarrow CO_2(aq) \; K_{H,CO_2} \tag{2-24}$$

The constant, K_H, for each of these reactions is the Henry's Law constant that equates the concentrations of the gases in the two phases, vapor and solution. Henry's Law constants are provided in reference works in many different units; however, the most convenient for groundwater purposes is moles/L·atm. If these units are used for K_H, then the unit for vapor phase concentration of the gas [e.g., $O_2(gas)$] is atmospheres and the unit for the solution phase concentration [$O_2(aq)$] is moles/L. The gas phase concentration is the partial pressure (in atmospheres) of the gas, which is simply that portion of the total pressure attributable to the individual gas. For example, oxygen comprises about 21% (on a mole basis) of the

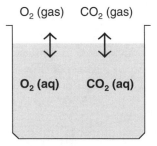

O_2 (gas) CO_2 (gas)

O_2 (aq) CO_2 (aq)

Figure 2-4   Equilibration of vapor phase and dissolved gases.

molecules in atmospheric air; therefore the partial pressure of oxygen in atmospheric air is 0.21 atmospheres. When using these units, the gas exchange equilibrium constant for oxygen can be written as follows:

$$K_{H,O_2} = \frac{O_2(aq)}{P_{O_2}} \qquad (2-25)$$

Henry's Law Constants for several important gases in groundwater are given in Table 2-4.

TABLE 2-4.

HENRY'S LAW CONSTANTS FOR GASES IN WATER (25°C)	
Gas	K_H **(moles L^{-1} atm^{-1})**
O_2	1.28×10^{-3}
CO_2	3.38×10^{-2}
H_2	7.9×10^{-4}
N_2	6.48×10^{-4}
CH_4	1.34×10^{-3}
NO	2.0×10^{-3}

From Manahan.[12]

The concentration of dissolved oxygen in water in equilibrium with atmospheric air can be calculated using Equation 2-25 and the partial pressure of oxygen in atmospheric air.

$$1.28 \times 10^{-3} \text{ moles/L atm} = \frac{O_2(aq)}{0.21 \text{ atm}}$$

$$O_2(aq) = (1.28 \times 10^{-3} \text{ moles/L atm})(0.21 \text{ atm})$$

$$- 2.69 \times 10^{-3} \text{ moles/L} \times 32 \text{ g/mole} \times 1000 \text{ mg/g}$$

$$= 8.6 \text{ mg/L } (25°C)$$

This is the solubility of molecular oxygen gas in water at 25°C under atmospheric conditions and represents the amount of dissolved oxygen expected to be initially present in water percolating into the subsurface. There are no reactions below the earth's surface that produce oxygen, however there are many types of reactions that consume oxygen. These include the oxidation of organic matter and redox reactions involving reduced metal species derived from the weathering of minerals:

$$CH_2O + O_2 \rightarrow CO_2 + H_2O$$

$$4Fe^{2+} + O_2 + 10H_2O \rightarrow Fe(OH)_3 + 8H^+$$

As dissolved oxygen is consumed by these types of reactions, additional oxygen will dissolve into the water from the vapor phase in the unsaturated zone, which in turn will lower the partial pressure of oxygen in this gas phase. Unless oxygen is resupplied from atmospheric air, the overall concentration of molecular oxygen in the unsaturated zone will decrease as a result of oxidation reactions. If the partial pressure of oxygen decreases, then the solubility of oxygen will also decrease. Below the water table the same type of oxygen consumption reactions will occur. Because there is no vapor phase below the water table, dissolved oxygen cannot be replenished and will be reduced to low levels if there is sufficient reactive, reduced material to be oxidized. In some cases such as near swamps or leaking petroleum storage tanks the amount of oxidizable material may be such that the dissolved oxygen level is reduced to very low levels near the water table (see Chapters 10 and 13 on landfills and organic contamination). In other instances, measurable dissolved oxygen may persist well below the water table if the aquifer material is not susceptible to oxidation.[13]

Solubility of Reactive Gases

The Henry's Law constant cannot be used to calculate directly the solubility of a gas if the gas produces other dissolved species in solution containing the component of the gas. For example, when CO_2 gas dissolves in water, it produces dissolved $CO_2(aq)$, which in turn reacts with the water to form the other inorganic carbon species — carbonic acid, bicarbonate, and carbonate. The solubility of the gas is the total amount of the gas that dissolves in the water and is equal to the sum of the inorganic carbon species in solution.

$$CO_2(g) \text{ solubility} = CO_2(aq) + H_2CO_3 + HCO_3^- + CO_3^{2-} \tag{2-26}$$

The $CO_2(aq)$ concentration is related to the $CO_2(g)$ partial pressure by its Henry's Law constant, and the other carbon species are related to each other by the formation constants (at 25°C) for each species.

$$CO_2(gas) \leftrightarrow CO_2(aq) \quad K_{H,CO_2} = 10^{-1.5} \tag{2-27}$$

$$CO_2(aq) + H_2O \leftrightarrow H_2CO_3 \quad K_{H_2CO_3} = 10^{-2.8} \tag{2-28}$$

$$H_2CO_3 \leftrightarrow HCO_3^- + H^+ \quad K_{HCO_3^-} = 10^{-3.5} \tag{2-29}$$

$$HCO_3^- \leftrightarrow CO_3^{2-} + H^+ \quad K_{CO_3^{2-}} = 10^{-10.3} \tag{2-30}$$

Equations 2-27 and 2-28 are usually combined to relate the carbon dioxide gas partial pressure to a term representing $CO_2(aq)$ plus H_2CO_3:

$$H_2CO_3^* = CO_2(aq) + H_2CO_3 \tag{2-32}$$

Using this equality,

$$CO_2(g) + H_2O \leftrightarrow H_2CO_3^* \quad K_{H_2CO_3^*} = 10^{-1.5} \tag{2-33}$$

$$H_2CO_3^* \leftrightarrow HCO_3^- + H^+ \quad K_{HCO_3^-} = 10^{-6.4} \times 10^{-8.4} \tag{2-34}$$

Figure 2-5 shows the dissolved concentrations of the three inorganic carbon species versus solution pH in a system open to atmospheric air, which has a carbon dioxide gas partial pressure of 0.0003 atm. As shown, the concentration of $H_2CO_3^*$ is constant at 10^{-5} moles/L at all pH values. This is reasonable because the formation reaction for $H_2CO_3^*$ (Eq. 2-33) does not include the hydrogen ion and therefore the concentration of $H_2CO_3^*$ is not pH dependent. The formation reaction for bicarbonate, Equation 2-34, does include the hydrogen ion, thus its concentration is pH dependent. As shown in Figure 2-5, the bicarbonate concentration versus pH increases with a slope of one. The bicarbonate concentration equals the $H_2CO_3^*$ at a pH value of 6.4, which is equal to the pK value of the formation reaction, Equation 2-34. Above a pH 6.4, the bicarbonate concentration exceeds the $H_2CO_3^*$ concentration. The carbonate (CO_3^{2-}) concentration is also pH dependent; however, because two hydrogen ions are involved in deriving carbonate from carbonic acid, the slope of the carbonate concentration line is two. Figure 2-5 shows that the carbonate concentration increases twice as fast as the bicarbonate concentration, and the two are equal at a pH of 10.3, which is the pK value of the formation reaction of carbonate from bicarbonate. The equations for the lines plotted in Figure 2-5 are derived from the formation reactions and are as follows:

$$\log H_2CO_3^* = -1.5 + \log\left(P_{CO_2}\right) = -1.5 + (-3.5) = -5$$

$$\log HCO_3^- = -7.9 + \log\left(P_{CO_2}\right) + pH = -11.4 + pH$$

$$\log CO_3^{2-} = -18.2 + \log\left(P_{CO_2}\right) + 2pH = -21.7 + 2pH$$

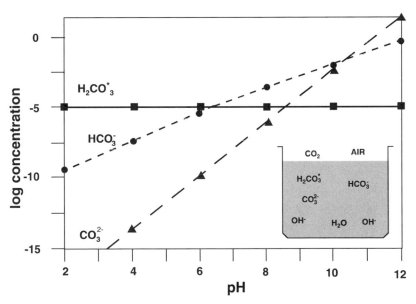

Figure 2-5   Carbonate species in an open system ($P_{CO_2} = 10^{-3.5}$ atm).

In summary, in a system open to exchange of CO_2 gas with a vapor phase, the relative proportion of the dissolved inorganic carbon species is pH dependent and the overall dissolved concentration of inorganic carbon increases as the pH increases. Therefore the solubility of carbon dioxide gas increases with pH. If speciation had been ignored in calculating the solubility of carbon dioxide gas in water and Henry's Law alone was used to calculate CO_2 solubility, the value would be 10^{-5} moles/L (0.4 mg/L as CO_2) at all pH values for a solution in equilibrium with atmospheric air. This value is only correct up to a pH of about 6. Above a pH of 6, the total inorganic carbon concentration is greater than 10^{-5} moles/L (Figure 2-5). For example, at a pH of 8 the solubility of CO_2 in water at a P_{CO_2} of 0.0003 atm is about 18 mg/L as CO_2.

Below the water table the aquifer is a closed system with respect to exchange with gases in a separate vapor phase. Assuming that there is no external source or sink for dissolved inorganic carbon in groundwater below the water table (that is, no carbonate minerals are present), the total dissolved inorganic carbon concentration will remain constant with pH change. However, ion speciation reactions still occur, and the dominant inorganic carbon species remain a function of the solution pH. This can be represented by the concentration diagram shown in Figure 2-6. This diagram is plotted assuming that groundwater at the water table was in equilibrium with the atmospheric air P_{CO_2} value of $10^{-3.5}$ atm. As in the open system, $H_2CO_3^*$ is the dominant species to a pH of 6.4, HCO_3^- dominates in the pH range of >6.4 to <10.3, and CO_3^{2-} is the dominant species above a pH of 10.3. The assumption that the system is closed to addition or loss of CO_2 gas is generally good in the saturated zone below the water table; however, the assumption that there are no sinks for the carbonate species does not generally hold. As the pH of groundwater increases due to weathering reactions, the concentration of the carbonate species (CO_3^{2-}) increases in the water and carbonate minerals become oversaturated and precipitate removing carbonate from solution. This reaction lowers the total inorganic carbon concentration in the groundwater while transferring it to the solid phase.

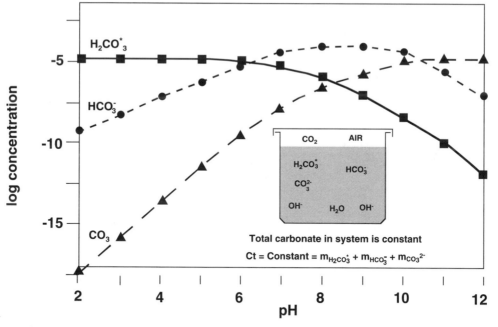

Figure 2-6   Carbonate species in a closed system (P_{CO_2} = $10^{-3.5}$ atm).

Carbon Dioxide Gas Pressures in the Subsurface

In the unsaturated zone, the partial pressure of CO_2 gas is often increased by factors of 100 or more compared with atmospheric air values as a result of the oxidation of organic matter. This process is shown in simplest form by the following reaction:

$$CH_2O + O_2 \leftrightarrow CO_2 + H_2O$$

This reaction consumes organic matter present in the soil and oxygen present in percolating water and in the pore spaces. It also raises the carbon dioxide gas level in the soil vapor. As the soil vapor partial pressure of CO_2 increases, the total amount of dissolved inorganic carbon also increases to maintain equilibrium between the gas and solution phases. For example, if the water in a closed system with respect to CO_2 is in equilibrium with a P_{CO_2} of $10^{-2.5}$ atm, then all of the concentrations of the dissolved species in Figure 2-6 would shift up by one log unit (a factor of ten) to reflect the higher dissolved concentrations. The presence of readily degradable organic matter in vegetation near the surface in the soil zone may account for much of the production of CO_2 gas in the subsurface. Beneath the soil zone in the deeper unsaturated regions, the transport of dissolved and particulate organic matter by recharge water may provide a continuing source of oxidizable carbon material to produce additional carbon dioxide gas.[14]

The fact that the inorganic carbon concentration in groundwater is a function of both solution pH and partial pressure of carbon dioxide gas in equilibrium with the water means that we can measure any two of these variables in a water sample and solve for the third. The relationship between these three properties is generally used to calculate the P_{CO2} pressure in equilibrium with groundwater using field-measured values for pH and alkalinity. The equation is as follows:

$$P_{CO_2}(atm) = \frac{\left(\gamma_{CO_3^{2-}}\right)\left(10^{-pH}\right)^2}{10^{-18.2}} \frac{Alkalinity\left(mg/L\,CaCO_3\right)}{\left(2 + \frac{10^{-pH}}{10^{-10.3}}\right)\left(5\times10^4\right)}$$

where $\gamma_{CO_3^{2-}}$ = activity coefficient of CO_3^{2-}, and alkalinity = total inorganic carbon alkalinity.

The activity coefficient of carbonate can be calculated from the Davies or Debye-Huckel equations, but is usually in the range of 0.6 to 1.0 for groundwater with a TDS less than 10,000 mg/L. This equation for P_{CO_2} can be used to determine whether a groundwater sample exposed to atmospheric air will degas and bubble off CO_2 into the atmosphere or dissolve CO_2 from the atmosphere into the water sample. Example 2-2 shows two calculations of P_{CO_2}. In one case, the calculated P_{CO_2} pressure for the groundwater is $10^{-1.5}$ atm. This P_{CO_2} is 100x the atmospheric air value of $10^{-3.5}$ atm; therefore the water sample will lose CO_2 gas to the atmosphere when exposed to surface conditions. The loss of CO_2 gas will allow the pH of the sample to rise because the carbonic acid concentration decreases as CO_2 is lost from the sample. The second example is a groundwater sample that is in equilibrium with a P_{CO_2} of only $10^{-4.3}$ atm. When this water is exposed to air, CO_2 will dissolve into it in order to equilibrate with the atmosphere. The pH of this sample will increase over time because the addition of dissolved CO_2 to the water increases the carbonic acid concentration. Equilibration of a groundwater sample with atmospheric air may have a large effect on the pH of that sample. Because pH is a master variable that impacts most of the other calculations done to understand the geochemistry of the system, it is important that as accurate an aquifer pH value as possible be measured and, consequently, that CO_2 gas exchange be held to a minimum. This is the reason that it is recommended that the pH of a water sample be measured in the field before it has interacted with atmospheric air (see Sections 9-2 and 9-5 on sampling techniques).

Example 2-2. Calculation of carbon dioxide partial gas pressures in equilibrium with groundwater.

2.2a: Alkalinity = 300 mg/L as $CaCO_3$, pH = 7.2, $\gamma_{CO_3^{2-}}$ = 1

$$P_{CO_2}(atm) = \frac{(1)(10^{-7.2})^2}{10^{-18.2}} \frac{(300)}{\left(2 + \frac{10^{-7.2}}{10^{-10.3}}\right)(5 \times 10^4)} \frac{(10^{-14.4})(300)}{(10^{-18.2})(10^{3.1})(5 \times 10^4)} = 10^{-1.5}$$

2.2b: Alkalinity = 25 mg/L as $CaCO_3$, pH = 8.9, $\gamma_{CO_3^{2-}} = 1$

$$P_{CO_2}(atm) = \frac{(1)(10^{-8.9})^2}{10^{-18.2}} \frac{(25)}{\left(2 + \frac{10^{-8.9}}{10^{-10.3}}\right)(5 \times 10^4)} \frac{(10^{-17.8})(25)}{(10^{-18.2})(10^{1.4})(5 \times 10^4)} = 10^{-4.3}$$

Oxygen and carbon dioxide are the two most important reactive gases in the subsurface. However, other reactive gasses may also be present. Under highly reducing conditions, such as found in the vicinity of swamps and other areas with highly reactive organic matter, methane (CH_4) and hydrogen sulfide (H_2S) gases may be present. The presence of these gases usually requires that the decomposition of the organic matter is facilitated by microorganisms. The methanogenic organisms facilitate reactions in which the organic carbon in a compound like acetate (HCH_3CO_2), which is initially present as C(O), is both oxidized to C(IV) in CO_2 and reduced to C(-IV) in CH_4. This reaction can be represented as follows:

$$HCH_3CO_2 \rightarrow CH_4 + CO_2$$

In the case of H_2S gas, the sulfur-reducing bacteria facilitate the following reaction:

$$2CH_2O + SO_4^{2-} + 2H^+ \rightarrow H_2S + 2CO_2 + 2H_2O$$

The presence of hydrogen sulfide gas can be noted by the characteristic rotten-egg odor. Its production is very significant in the evaluation of metal mobility because many heavy metals form very insoluble sulfide species under reducing conditions.

These reactions, which produce methane, hydrogen sulfide, and carbon dioxide gasses and consume oxygen, may also occur in the subsurface near landfills and leaks or spills of organic compounds such as hydrocarbons. It has been shown[15] that the distribution and concentrations of gasses such as CO_2, O_2, CH_4, and H_2S in the unsaturated zone may be used to localize the source of organic contaminants in the subsurface.

Hydrogen is another reactive gas produced under very reducing conditions. It does not usually occur at very high concentrations under natural or even most contaminant situations; however, it has been produced at explosive levels in a well bore during percussion drilling of alluvial material.[16] Hydrogen was apparently generated under these conditions as a result of a reaction between the steel drive casing, the sediments, and the groundwater.

REFERENCES

1. Truesdell, A., The advantage of using pe rather than Eh in redox equilibrium calculations, *J. Geol. Ed.,* 17, 17–20, 1969.
2. Lindsay, W.L., *Chemical Equilibria in Soils,* John Wiley, New York, 1979, pp. 1–449.
3. Hart, E.J., Sheffield, G., and Fielden, E.M., Reaction of the hydrated electron with water, *J. Phys. Chem.,* 70, 150–156, 1966.

4. Lindberg, R.D. and Runnells, D.D., Ground water redox reactions: an analysis of equilibrium state applied to Eh measurements and geochemical modelling, *Science*, 225, 925–927, 1984.
5. Whitfield, M., Thermodynamic limitations on the use of the platinum electrode in Eh measurements, *Limnol. Oceanogr.*, 19, 857–865, 1974.
6. Nordstrom, D.K., Jenne, E.A., and Ball, J.W., in *Chemical Modeling in Aqueous Systems*, Jenne, E.A., Eds., American Chemical Society, 1979.
7. Beaucaire C. and Toulhoat, P., Redox chemistry of uranium and iron, radium geochemistry, and uranium isotopes in the groundwaters of the Lodeve Basin, Massif Central, France, *Appl. Geochem.*, 2, 417–426, 1987.
8. Davis, A. and Ashenberg, D., The aqueous geochemistry of the Berkeley Pit, Butte, Montana, U.S.A., *Appl. Geochem.*, 4, 23–36, 1989.
9. Walton-Day, K., Macalady, D.L., Brooks, M.H., and Tate, V.T., Field methods for measurement of ground water redox chemical parameters, *GWMR*, 81–89, 1990.
10. Stollenwerk, K.G., Geochemical interactions between constituents in acidic groundwater and alluvium in an aquifer near Globe, Arizona, *Appl. Geochem.*, 9, 353–369, 1994.
11. Cherry, J.A., Shaikh, A.U., Tallman, D.E., and Nicholson, R.V., Arsenic species as an indicator of redox conditions in groundwater, *J. Hydrol.*, 43, 373–392, 1979.
12. Manahan, S.E., *Environmental Chemistry*, Lewis, New York, 1991.
13. Winograd, I.J. and Robertson, F.N., Deep oxygenated ground water: anomaly or common occurrence?, *Science*, 216, 1227–1230, 1982.
14. Wood, W.W., Origin of caves and other solution openings in the unsaturated (vadose) zone of carbonate rocks: a model for CO_2 generation, *Geology*, 13, 822–824, 1985.
15. Robbins, G.A., McAninch, B.E., Gavas, F.M., and Ellis, P.M., An evaluation of soil-gas surveying for H_2S for locating subsurface hydrocarbon contamination, *GWMR*, 124–132, 1995.
16. Bjornstad, B.N. et al., Generation of hydrogen gas as a result of drilling within the saturated zone, *GWMR*, 140–147, 1994.

3 WATER/ROCK INTERACTIONS

This chapter focuses on the interactions that occur between groundwater and the solid phases comprising the unsaturated and saturated zones. As discussed previously (Section 1.3), these solid phases consist of inorganic material (minerals and amorphous compounds) and organic matter. Solids are important to the geochemistry of the system because they are the primary sources and sinks of dissolved constituents. They may limit the dissolved concentration of a solute if the solubility of a mineral containing the solute is reached. However, even if mineral precipitation does not limit solution concentration, other water/rock process such as adsorption/desorption will affect the mobility of most dissolved constituents. In addition, solids are the primary contributors to the acid/base neutralizing capacity and oxidizing/reducing capacity of the system. The primary water/rock interactions described in this chapter are adsorption/desorption and mineral precipitation/dissolution processes.

3.1. ADSORPTION/DESORPTION

Adsorption is the removal of a dissolved species from solution by its attachment to the surface of a solid. Desorption is the release of the species back into the solution. If the solid is immobile, adsorption temporarily removes some of the solute from the mobile solution phase and slows down the movement of the mass of the adsorbing species relative to the groundwater flow rate. If the adsorbing solid is a colloidal particle suspended in the solution, then adsorption will not retard movement of the solute as long as the solid remains suspended in the flow; however, adsorption still lowers the effective concentration, activity, of the solute. In this section adsorption will be discussed in terms of (1) ion exchange, which primarily impacts the major cations in solution; (2) adsorption isotherms and distribution coefficients, which provide lumped parameters that equate adsorption of trace elements onto soil material; and (3) surface complexation, which describes the mechanism of adsorption onto specific solid phases and also provides a method of quantifying adsorption.

Ion Exchange

Ion exchange is a type of adsorption/desorption phenomenon that applies principally to material with a porous lattice containing fixed charges. Clay minerals are the most common ion exchangers in the soil and aquifer environments. The fixed charge on clay minerals is a result of the substitution of Al^{3+} for Si^{4+} in the tetrahedral clay lattice sites and Fe^{2+} and Mg^{2+} for Al^{3+} in the octahedral lattice sites. The presence of the lower charged cation in the structure results in a net negative charge on the surface of the clay (Figure 3-1). Electrostatic attraction brings dissolved cations to the clay surface to balance the charge.

The exchange capacity (number of available exchange sites) of silicate clay minerals is a function of the amount of substitution that has occurred in the crystal lattice of the mineral and the amount of available surface area. As shown in Table 3-1, montmorillonite has the highest amount of substitution and external/internal surface area, which is reflected in the greatest cation exchange capacity. Illite has less substitution than montmorillonite and its surfaces are strongly held together by K^+ ions resulting in low available internal

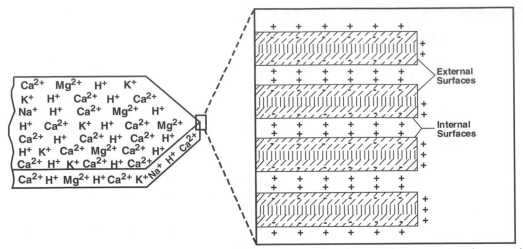

Figure 3-1   Schematic of silicate clay mineral showing internal and external net negative charge with attached exchangeable cations. (From Brady, N.C., *The Nature and Properties of Soils*, Macmillan, New York, 1974, with permission.)

TABLE 3-1.

CLAY PROPERTIES IMPACTING CATION EXCHANGE CAPACITY			
Property	**Montmorillonite**	**Illite**	**Kaolinite**
Size (μm)	0.01–1.0	0.1–2.0	0.1–5.0
Ion substitution	High	Medium	Low
Surface area (m^2/g)	600–800	100–120	5–20
External surface	High	Medium	Low
Internal surface	Very high	Medium	None
Cation exchange capacity (meq/100 g)	80–100	15-40	3–15

surface area and smaller exchange capacity than montmorillonite. Kaolinite has the least amount of substitution and the lowest surface area and cation exchange capacity.

Although most of the cation exchange capacity of clay minerals is due to the permanent fixed charge at the clay surface, there is also a pH-dependent charge that can contribute to the exchange capacity. At low pH values (less than about 6) hydrogen ions are strongly bonded to oxygen atoms at crystal edges, and these sites are not available for cation exchange. However, as the pH increases and the aqueous hydrogen ion activity decreases, the bond with oxygen breaks releasing hydrogen into solution, thereby creating new cation exchange sites. For montmorillonite, the amount of cation exchange capacity attributed to the pH-dependent charge may be 10% of the total CEC.[1]

Organic matter (primarily natural humic substances) provides the other potentially important source of ion exchange capacity in the environment. The major source of negative charge on the surface of humus is exposed carboxylic (–COOH) and phenolic ($\bigcirc$–OH) groups associated with a central organic solid unit (Figure 3-2). The charge on the humic surface is pH dependent in a manner similar to that of part of the charge on the silicate clay minerals. At low pH values, H$^+$ is the major adsorbing, exchangeable cation, but as the pH increases and the H$^+$ activity decreases, the sites become more available for exchange with the major cations in solution. The overall cation exchange capacity of humus is on

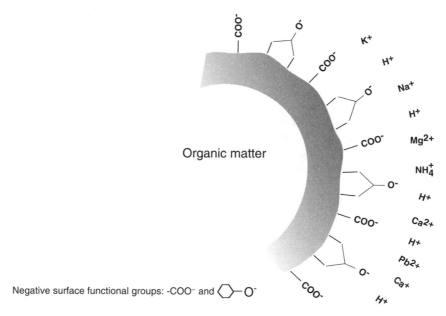

Figure 3-2 Schematic of cation adsorption onto organic matter. (Note that adsorption occurs on internal surfaces, as well as the external surface shown.)

the order of 200 meq/100 g, which is significantly greater than that of the silicate clay minerals.[1]

The cation exchange capacity available to soil water or groundwater will be a function of the amount of inorganic clay minerals and organic humus present in the solid phase. Material coarser than clay size (2 μm) has a proportionally smaller surface area and does not add significantly to the cation exchange capacity. Appelo and Postma[2] provide the following empirical formula that equates CEC to the concentrations of clay minerals and organic matter in the solid phase:

$$\text{CEC} \left(\text{meq}/100\,\text{g} \right) = 0.7 \times \left(\% \text{ clay} \right) + 3.5 \times \left(\% \text{ C} \right)$$

This formula will hold for an average mix of silicate clay minerals in the solid phase. Because the CEC of individual clay minerals varies so greatly (Table 3-1), a material that contains predominantly kaolinite will have a clay factor less than 0.7 and one that is predominantly montmorillonite will have a greater clay factor.

Anion exchange may also occur on clay minerals, but to a much lesser extent than cation exchange because of the dominant fixed negative charge on the clay mineral surface. However, the adsorption of H^+ onto the surface can produce a positively charged exchange site according to the following reaction:

$$\equiv MOH + H^+ \leftrightarrow \equiv MOH_2^+ \qquad\qquad (3\text{-}1)$$

where $\equiv M$ is a metal ion in the mineral lattice. The positively charged sites are available for anion (A) exchange as shown by:

$$\equiv MOH_2^+ + A^- \leftrightarrow \equiv MOH_2A \qquad\qquad (3\text{-}2)$$

In addition, anion exchange may occur by displacement of the hydroxyl species as shown by the reaction:

$$\equiv MOH + A^- \leftrightarrow \equiv MA + OH^- \qquad (3\text{-}3)$$

The anion exchange capacity of clays is usually negligible at the pH of most natural material; however, it can be on the order of 10% of the CEC for more heavily weathered soils where conditions are relatively acidic.[3]

Because ion exchange is primarily an electrostatic process, the more highly charged solution species are preferentially adsorbed. When present at equal solution concentrations, the affinity of the solid exchanger for solutes is as follows:

$$Al^{3+} > Ca^{2+} > Mg^{2+} > K^+ > Na^+$$

In addition to the charge on the ion, the smaller the radius of the ion, the closer it can approach the surface and the greater the affinity of the surface for the ion. However, ions in solution are surrounded by water molecules; therefore their "effective" radius is the radius of the solvated ion and not the bare ion. The larger the bare ion, the smaller the solvated radius; thus for the monovalent ions the affinity of the solid exchanger is as follows:

$$Cs^+ > Rb^+ > K^+ > Na^+ > Li^+$$

This order of selectivity (known as the Lyotropic series) assumes that the ions are present at equal concentrations in solution. To calculate the distribution of ions between the solution and solid phase, ion exchange reactions must be written with an associated equilibrium constant. In the case of Na^+ and K^+ exchange the reaction can be written as follows:

$$X-K + Na^+ \leftrightarrow X-Na + K^+ \qquad (3\text{-}4)$$

where X is the solid exchanger. For this reaction the equilibrium constant is as follows:

$$K_{Na/K} = \frac{\left(a_{X-Na}\right)\left(a_{K^+}\right)}{\left(a_{X-K}\right)\left(a_{Na^+}\right)}$$

Although we have a good method of calculating the activities of dissolved solutes from measured solution concentrations (Section 1.3), there is not an equivalent method for calculating the activities of the adsorbed species. Appelo and Postma[2] provide a good review of the conventions employed to represent the adsorbed fraction and calculate a K value. It turns out that, given the available methods of calculating the equilibrium constants for ion exchange reactions, the K is not very constant. K varies with pH because of the pH-dependent charge on the solid surfaces, and it varies with total solute concentrations when heterovalent ions (such as Na^+ and Ca^{2+}) are exchanging. Furthermore, K varies with the type of exchanger present in the solid phase. For these reasons, the K values for ion exchange reactions may better be considered exchange coefficients instead of equilibrium constants. Appelo and Postma[2] compiled exchange coefficients for several major and minor ions in groundwater. Their data can be used to expand the selectivity series for the divalent ions to the following:

$$Pb^{2+} > Ba^{2+} \approx Sr^{2+} > Cd^{2+} \approx Zn^{2+} \approx Ca^{2+} > Mg^{2+} \approx Ni^{2+} \approx Cu^{2+} > Mn^{2+} > Fe^{2+} \approx Co^{2+}$$

The exchange coefficients for all of these divalent ions are fairly similar with the coefficient for the most strongly attracted ion (Pb^{2+}) only a factor of two greater than the coefficient for the least strongly attracted (Fe^{2+} and Co^{2+}). In typical $Ca-HCO_3^-$ groundwater, Ca^{2+} and Mg^{2+} are the dominant ions in solution, and their relatively high concentration will result in their being the dominant ions on the exchange sites even though some of the trace metals may be more strongly attracted to the exchange sites. However, in contaminant situations where one of the contaminant ions that is more strongly attracted to the exchange sites than Ca^{2+} or Mg^{2+}, such as lead or barium, is present at similar concentrations as the normal major ions, then the exchange sites may preferentially hold the contaminant and affect its mobility in the aquifer.

Perhaps the most common impact of ion exchange on water quality is the case of seawater intrusion into a freshwater aquifer.[2,4,5] As freshwater is pumped from an aquifer near a coastline, seawater will flow toward the well to partially compensate for drawdown of the water table. The initial ion exchange sites will be predominantly occupied by Ca^{2+} and Mg^{2+} because of their greater valence and concentration in the groundwater than Na^+ and K^+. Seawater is predominantly a NaCl solution, and the high sodium concentration in the intruding seawater will drive the ion exchange reaction in the direction to adsorb Na^+ and release the divalent cations.

$$Na^+ + 1/2\,X_2-Ca \rightarrow X-Na + 1/2\,Ca^{2+} \tag{3-5}$$

The ion exchange process retards the movement of Na^+ relative to an ion such as Cl^- in the seawater that is not as strongly affected by ion exchange or other water/rock interactions.

Distribution Coefficients

Although the trace constituents in groundwater may not be the dominant exchangeable species because they cannot successfully compete for the exchange sites with the major ions, they will participate in adsorption reactions and will come to some equilibrium between solution and adsorbed concentrations given sufficient residence time. These reactions with the solid surfaces of the system will impact the mobility of the trace constituents.

The degree of trace element adsorption is studied in the laboratory by beaker experiments in which aquifer material is mixed with groundwater containing adsorbable species (Figure 3-3). In these experiments, groundwater with a known concentration of the dissolved species is brought into contact with the adsorbent aquifer material for a period of time in which it is believed that equilibrium will be achieved between the dissolved and adsorbed concentrations of the species. The final solution concentration is measured, and the difference between initial and final solution concentration is used to calculate the amount adsorbed. The reaction can be written as follows:

$$X + C_{aq} \leftrightarrow C_{ads} \tag{3-6}$$

In this reaction the solid adsorbent is represented as X, the dissolved species is C_{aq}, and the adsorbed species on X is shown as C_{ads}. The relationship between these three constituents in the reaction is generally referred to as a distribution (Kd) (or partition (Kp)) coefficient to reflect the fact that the species is distributed or partitioned between the solution and the solid phases.

$$Kd\,(or\,Kp) = \frac{(C_{ads})}{(X)(C_{aq})} \tag{3-7}$$

The Kd is an empirically derived constant and is not a thermodynamic equilibrium constant as are the constants for the speciation and redox reactions previously considered. Note also that the adsorption reaction (Eq. 3-6) used for the distribution coefficient does not involve ion exchange. Although there will be competition for the surface sites and ion exchange may be occurring, this part of the reaction is not incorporated in the defining reaction for the distribution coefficient. Finally, the Kd is a lumped parameter that incorporates all the possible types of surface adsorption processes that might remove species from the solution. It does not equate adsorption to a particular process such as attraction to a clay mineral surface or complexation with a surface functional group. It can also include precipitation of the species in a solid phase.

If we assume that the amount of adsorbent (X) is very large compared with the amount of available species, then adsorption does not significantly change the concentration of surface sites (X) in Equation 3-7, and X can be set equal to 1. This simplifies the Kd equation to the following:

$$\text{Kd} = \frac{C_{ads}}{C_{aq}} \quad \left(\text{Units: } \frac{mg/kg}{mg/L} = L/kg \right) \tag{3-8}$$

The quantity (C_{ads}) is the concentration of the contaminant adsorbed on the solid; therefore it has units of milligrams of C per kilogram of solid. The quantity (C_{aq}) is the solution concentration of the adsorbed species, and it has units of milligrams of C per liter of solution. Consequently, Kd has units of L/kg.

Figure 3-3 shows the results of several beaker experiments in which different starting solution concentrations of the contaminant were contacted with the solid. A plot of final dissolved concentration versus adsorbed concentration is a straight line in which each point on the line represents the results of a separate batch experiment. The line is referred to as an isotherm because it is developed at constant temperature, and the Kd is called a linear adsorption isotherm because it plots as a straight line. The Kd for this species on the solid used in the experiment is the y-value (C_{ads}) divided by the x-value (C_{aq}) at any point on the line. K_d also equals the slope of the line because the line passes through the origin.

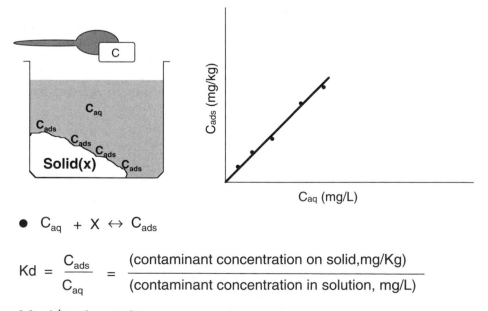

Figure 3-3 Adsorption reactions.

Two separate linear isotherms are plotted on Figure 3-4 to represent the adsorption of two different species on a solid. The difference in slopes of the lines represents the different affinities of the surface for the two species. (The two isotherms could also represent the different affinities of two different solids for the same dissolved species.) The isotherm with the greater slope (closer to the y-axis, C_{ads}) represents a solid that strongly attracts the contaminant to its surface so that the calculated Kd is large. A high Kd reflects a preference for the species to be on the solid rather than dissolved in the solution. The isotherm with the lower Kd (near the x-axis, C_{aqs}) reflects the relative preference of the species to stay in solution rather than be adsorbed. The Kd value is a measure of the relative affinity of the solid phase for the two species.

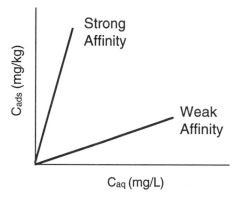

Figure 3-4 Adsorption affinity of solid.

A straight line plot of adsorption implies that the process of adsorption is not affected by solute concentration and that the surface of the solid has unlimited capacity for adsorption. Linear adsorption isotherms are appropriate for many species at low solution concentration, but they are not appropriate at higher concentrations when the surface sites for adsorption become filled. The three general types of isotherms produced by adsorption experiments are shown in Figure 3-5. The adsorption response curves represent the linear isotherm, the Freundlich isotherm, and the Langmuir isotherm. The relationships between dissolved and adsorbed concentrations for the Freundlich and Langmuir isotherms are as follows:

$$C_{ads} = K_F \left(C_{aq} \right)^{1/n} \tag{3-9}$$

$$C_{ads} = \frac{K_L A_m C_{aq}}{1 + K_L C_{aq}} \tag{3-10}$$

where n = a fitting parameter for the curve of the Freundlich isotherm, unitless, and A_m = adsorption capacity term for the Langmuir isotherm, mg/kg.

Each of the isotherms is linear at low adsorbate concentration; however, the Freundlich and Langmuir isotherms change slope at higher concentrations. The Freundlich isotherm becomes a curve at higher concentration reflecting lower adsorption at these values as the adsorption sites become filled. However, there is no total capacity term in the Freundlich isotherm equation, so there is no upper limit on adsorption. The Langmuir isotherm does have a capacity term (A_m) in its definition. Once the concentration of adsorbed species reaches this capacity term, adsorption decreases to zero and any additional increase in species solution concentration remains in solution.

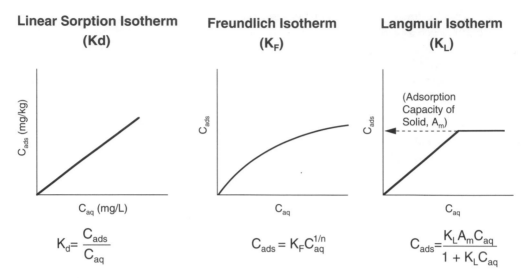

Figure 3-5   Adsorption response curves.

In the linear range of an isotherm, the Kd can be calculated from a single batch experiment using the final values for concentrations in solution and on the solid. If adsorption response follows the Freundlich or Langmuir isotherms, then several experiments must be conducted to determine the shape of the curve. To determine the K_F and n values for a Freundlich isotherm, the experimental data are plotted with the log C_{aq} versus log C_{ads}. If the data plot as a line using these axes, then log K_F is the y-intercept and 1/n is the slope of the line. To determine the K_L and A_m values when adsorption follows a Langmuir curve, the data are plotted with the y-axis equal to C_{aq}/C_{ads} and the x-axis equal to C_{aq}. If the data plot as a line using these axes, then the y-intercept is $1/K_L A_m$ and the slope is $1/A_m$.

The adsorption of organic solutes onto soil and aquifer solids can also be represented by a distribution coefficient. The dissolved hydrophobic compounds, which would much rather be associated with solid organic compounds than be dissolved in water, will adsorb onto the naturally occurring organic matter in the subsurface. The distribution coefficient for this adsorption is represented as follows:

$$Kd = k_{oc} f_{oc}$$

where K_{oc} is the organic carbon partition coefficient for the compound of interest and f_{oc} is the fraction of organic carbon present along the flow path. This relationship holds when f_{oc} is generally greater than 0.001. At lower f_{oc} values, inorganic compounds present in the aquifer may play a significant role in adsorbing organic compounds. The K_{oc} for a compound is estimated from (1) beaker tests in which octanol is used as a substitute for organic carbon in the aquifer, (2) the solubility in water of the compound, (3) beaker tests with aquifer solids (similar to the tests with trace inorganic compounds), and (4) field data. The octanol water partition coefficients (K_{ow}s) have been measured and compiled for many organic compounds.[6-8] They can be used with the following relationships published by the U.S. EPA[6] to calculate a K_{oc}:

semivolatile, nonionizing organic compounds:

$$\log K_{oc} = 0.00028 + (0.983 \log K_{ow})$$

volatile, nonionizing organic compounds:

$$\log K_{oc} = 0.0784 + (0.7919 \log K_{ow})$$

The practical outcome of adsorption and desorption is that a dissolved contaminant added to an aquifer will be retarded in its downgradient movement relative to the flow of groundwater. In other words, the velocity of the contaminant (v_c) will be less than the groundwater velocity (v_{gw}). Part of the appeal in developing adsorption isotherms is the ease of utilizing the adsorption constant to calculate retardation of an adsorbing species. The amount of retardation will be a function of the affinity of the solid for the species as reflected in the appropriate sorption coefficient. The following retardation factors, R = v_{gw}/v_c, have been developed for each isotherm[9]:

$$R = 1 + (\rho_\beta/\eta)Kd \quad \text{(linear isotherm)} \tag{3-11}$$

$$R = 1 + (\rho_\beta/\eta)(K_F n)(C_{aq}^{n-1}) \quad \text{(Freundlich isotherm)} \tag{3-12}$$

$$R = 1 + (\rho_\beta/\eta)\left[K_L A_m/(1 + K_L C_{aq})^2\right] \quad \text{(Langmuir isotherm)} \tag{3-13}$$

In these equations for contaminant retardation, ρ_β is the bulk density of the aquifer and η is its porosity. The ratio ρ_β/η reflects the mass of aquifer material in contact with one liter of groundwater. Because the units of ρ_β/η are g/cc (= kg/L), when this ratio is multiplied by Kd (units of L/kg), a unitless number is derived for the retardation factor. Kd is the ratio of the concentration on the solid to the concentration in the solution while R is the ratio of the absolute mass on the solid to the absolute mass in solution. Note that as the contaminant solution concentration (C_{aq}) increases, retardation decreases for the Freundlich (when n < 1) and Langmuir isotherms. This is to be expected as the adsorption sites on the solid become filled.

For typical porous media, the bulk density ranges from 1.6 to 2.1 g/cc and the porosity ranges from 0.2 to 0.4.[10] In Table 3-2, these values are used with representative Kd values to calculate retardation factors. It can be seen that even fairly small Kd values (<1) can have an important effect on the mobility of a contaminant and that higher Kd values can effectively immobilize a contaminant unless flow velocities are very high. It should be kept in mind that the adsorption capacity of the aquifer should also be considered in evaluating contaminant mobility, and that at high concentrations the response may not be linear and the mobility may be greater than would be the case for a linear response. Contaminant mobility is discussed further in Chapter 7.

Adsorption isotherms are developed for specific trace constituents under particular site conditions that include a set concentration of adsorbing surfaces and fixed pH, Eh, and concentrations of competing ions. If any of the conditions are altered, the adsorption isotherm parameters will likely change. Because the technique for determining the parameters does not provide information on the adsorption mechanism, it is not possible to estimate the effect of changing conditions on the parameters to any degree of accuracy. The surface complexation approach to adsorption provides a method of estimating adsorption affinity and capacity over a wide range of environmental conditions.

Surface Complexation Processes

Surface complexation is the attachment of species to the functional groups present on the solid surfaces of amorphous aluminosilicates, metal oxides/hydroxides and organic matter. Although ion exchange is a type of surface complexation, it is generally restricted to cation

TABLE 3-2.

REPRESENTATIVE RETARDATION FACTORS

Kd (L/kg)	Rf
0.25	2–3.5
0.5	3–6
1	5–11
2	9–21
10	41–106
100	401–1050

Note: Rf = 1 + (ρ_β/η)Kd; bulk density (ρ_β) = 1.6 to 2.1 g/cc; porosity (η) = 0.2 to 0.4.

exchange associated with the permanent charge on aluminosilicate clay mineral surfaces. Ion exchange reactions occur primarily in response to electrostatic attraction and are characterized by an exchange (selectivity) coefficient, while surface complexation reactions occur in response to both chemical and electrostatic components. Surface complexation is the mechanism by which trace elements are removed from solution, and it produces the adsorption response shown by plotting the various isotherms described above. However, unlike the isotherm approach to adsorption, surface complexation theory provides a method of predicting adsorption response over a wide range of environmental conditions (pH, Eh, and solution concentration).

Surface complexation is analogous to aqueous complexation, which is the close association of a central solute molecule surrounded by other dissolved solutes, such as carbonate complexing with hydrogen ions to form the bicarbonate (HCO_3^-) and carbonic acid (H_2CO_3) species. In surface complexation, the functional groups on the surface of the solid attract solutes, which complex with the attached functional group. Just as any and all solutes can form aqueous complexes to some degree, the surfaces of the solids will have some attraction for all the dissolved species. The degree of surface complexation for any particular species will depend on the chemical affinity of the surface for the species and the electrostatic (physical) attraction or repulsion of the species by the charged solid surface. The overall adsorption constant (K_{ads}) for the surface complexation reaction can be represented as follows:

$$K_{ads} = K_{int} \times K_{coul} \tag{3-14}$$

where K_{int} is the intrinsic chemical component of the adsorption constant and K_{coul} is the electrostatic (coulombic) component. The electrostatic component of the adsorption equilibrium constant arises from the amount of work that must be accomplished to bring the solute to the surface of the solid by passing through the potential gradient present near the solid surface due to the charge on the surface:[11]

$$K_{coul} = \exp(-\Delta ZF\Psi/RT) \tag{3-15}$$

where ΔZ is the change in the charge of the surface species due to the adsorption reaction, F is Faraday's constant, Ψ is the surface potential, R is the gas constant, and T is the absolute temperature. Substituting for K_{coul} in Equation 3-14, the adsorption equilibrium constant can be written as follows:

$$K_{ads} = K_{int} \times \exp(-\Delta ZF\Psi/RT) \tag{3-16}$$

The determination of K_{ads} by adjusting the intrinsic (chemical) component of the equilibrium constant by a factor $(exp(-\Delta ZF\Psi/RT))$ related to the charge of the complexing species is analogous to aqueous complexation in which the equilibrium constant based on concentrations of the free species (chemical component of K) is adjusted by the activity coefficient for the effect of physical attraction or repulsion of the ions. In fact, the calculation of the electrostatic interaction term for surface complexation based on the Gouy-Chapman theory is formally and conceptually similar to the Debye-Huckel theory of nonideal behavior of ions in solution.[12]

Surface complexation is not as straightforward as aqueous complexation because of the changing electrostatic characteristics of the solid surface as solutes are adsorbed to it. The adsorption of a charged particle may change the charge on the surface, which, in turn, will change the surface potential, Ψ, in the K_{ads} equation. In addition, because the adsorption sites are fixed on the surface of the solid and have a particular geometric arrangement, the presence of a particular species at one site will affect the adsorption of other ions of that species at neighboring sites. For example, if most of the surface sites contain hydrogen ions, it will be more difficult to adsorb additional hydrogen ions relative to a similar monovalent ion because of the repulsion between adjacent hydrogen ions on the surface.

Figure 3-6 is a model of a solid with available adsorption sites that carry a charge because the metal (M) and oxygen (O) atoms on the surface are not fully coordinated with balancing atoms as are the atoms beneath the surface of the solid. In water, hydrogen and hydroxyl ions will be attracted to the surface and form surface complexes with the solid. In contrast to clay minerals, which have a relatively small component of pH-dependent charge, the charge on amorphous aluminosilicates, metal oxides/hydroxides, and organic matter is predominantly pH dependent. As the pH changes, charge is developed by (1) complexation of H^+ with surface OH^- groups, (2) stripping of hydrogen from surface OH^- groups, and (3) loss of OH^- groups from the surface as shown in Figure 3-6 and the following reactions:

$$\equiv M-OH + H^+ \leftrightarrow \equiv MOH_2^+ \tag{3-17}$$

$$\equiv M-OH \leftrightarrow \equiv MO^- + H^+ \tag{3-18}$$

$$\equiv M-OH \leftrightarrow \equiv M^+ + OH^- \tag{3-19}$$

At lower pH values, the surface is more positively charged because of the presence of additional complexed hydrogen ions (Eq. 3-17) producing $\equiv MOH_2^+$ and the loss of OH^- from the surface (Eq. 3-19) which leaves an uncompensated positive charge ($\equiv M^+$). At higher pH values, the surface is predominately negatively charged due to the loss of H^+ from the surface (Eq. 3-18). Because of the predominance of positively charged surface sites at lower pH values, the solid preferentially adsorbs anions under these conditions (Figure 3-7a), while at higher pH values the predominantly negatively charged surface becomes more attractive to cations (Figure 3-7b). Note that the layer with adsorbed cations or anions may itself carry a charge because adsorption has not completely neutralized the charge of the adsorbing species. Because the charge on this layer is generally less than that on the solid surface, electrostatic attraction to this layer is reduced and the charge is balanced by a diffuse layer of counter ions in the solution (Figure 3-7).

The pH at which the concentration of positively charged surface sites equals the concentration of negatively charged surface sites is called the pristine point of zero charge, or PPZC.[13] At the PPZC the $\equiv MOH_2^+$ surface charge is balanced by the $\equiv MO^-$ charge minimizing the electrostatic attraction of other cations and anions. At pH values below the

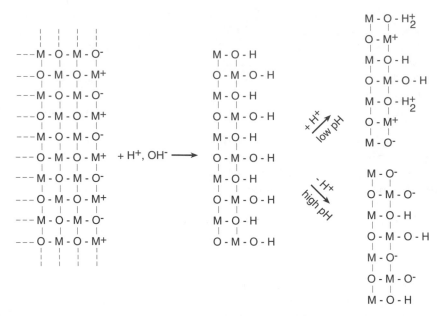

Figure 3-6   Model of solid adsorbent and surface complexation with H⁺ and OH⁻.

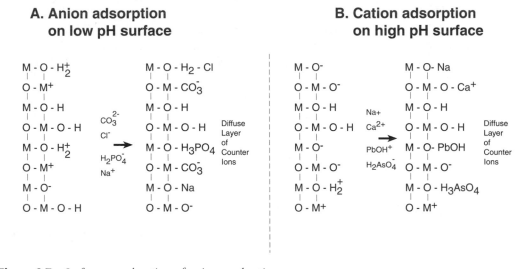

Figure 3-7   Surface complexation of anions and cations.

PPZC, the net surface charge is positive and anion adsorption is favored; while at pH values above the PPZC the net charge is negative and cation adsorption predominates. Table 3-3 provides the PPZCs for several commonly occurring minerals. Note that only the net surface charge is positive or negative depending on the pH and that some oppositely charged surface sites will be available at each pH.

The pH effect on adsorption can be seen in experimental data when the amount of a trace cation or anion adsorbed is plotted versus pH. As shown in Figure 3-8, as the pH increases the amount of cation adsorption increases in response to the increasing number of available negatively charged surface sites ($\equiv M-O^-$). For anions, adsorption increases as the pH decreases because more of the surface sites are positively charged ($\equiv MOH_2^+$ and $\equiv M^+$) at lower pH values. The location where adsorption increases significantly for a cation or anion is know as the "pH edge." The edge moves to higher pH values for

TABLE 3-3.

PRISTINE POINT OF ZERO CHARGE FOR REPRESENTATIVE MINERALS	
Mineral	**Point of zero charge**
Quartz	2–3
Feldspar	2–2.4
Montmorillonite	2.5
Kaolinite	4–5
Goethite	7–8
Ferrihydrite	7.5–8
Gibbsite	8–9

cations as the solution concentration of the adsorbing species (sorbate) increases or the amount of sorbent decreases primarily because of saturation of available sites. As the pH increases, more negatively charged sites develop on the surface so that a proportionately higher percentage of the cation can be adsorbed. The same relationship pertains to anions where more positively charged sites become available as the pH is lowered. The anion data also show a plateauing of adsorption at increasing sorbate/sorbent ratios. This is due to total saturation of the sites. The plateau effect is not generally seen with cations because at high pH values precipitation reactions will also limit cation concentrations in solution, and the percent of the cation adsorbed will include the amount due to surface complexation and precipitation.[11]

The pH edge generally does not occur at the point of zero charge of the solid. For cations the pH edge is below the point of zero charge and for anions it is above. The reason for this is that the pH edge is only the boundary for the predominance of one charge over another and cation adsorption sites are available below the point of zero charge, just as anion sites are available above the PPZC. Also, as shown by the equation for K_{ads} (3-14), the electrostatic term in only part of the adsorption equilibrium constant, and chemical affinity, as represented by the intrinsic adsorption constant (K_{int}), must also be taken into account in determining the amount adsorbed. The chemical affinity of a solid for a solute may be strong enough to allow for adsorption even though there is electrostatic

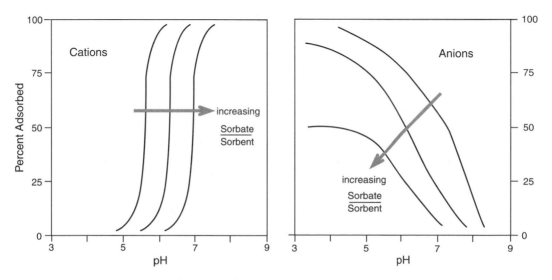

Figure 3-8   Cation and anion adsorption edges.

repulsion of the solute by the surface. The differing chemical affinities of the solid surface for the solutes produce specific adsorption effects in which certain species are preferentially removed from solution and held by the surface sites. For example, under low pH conditions where most of the sites may be initially filled by hydrogen ions ($\equiv MOH_2^+$), the addition of Pb^{2+} to the system may lead to Pb^{2+} adsorption and release of the hydrogen ions if the surface site has a stronger affinity for Pb^{2+} than H^+. The amount of Pb^{2+} adsorption and H^+ desorption will also be dependent on the relative activities of Pb^{2+} and H^+ in solution.

Part of the chemical affinity is due to the ability of the adsorbing molecule to "fit" among the tightly packed molecules that make up the surface of the solid. For example, PO_4^{3-} and AsO_4^{3-} appear to be strongly adsorbed onto the surface of iron oxyhydroxides. This is analogous to the good fit of K^+ between clay layers. Once adsorbed, these molecules may not be readily exchangeable with the solution, which makes determining the K_{ads} for the reaction difficult.

The intrinsic adsorption constants (chemical component of K_{ads}) for several cations and anions onto hydrous ferric oxide are presented in Table 3-4. These data were derived from experimental data or estimated from linear free-energy relationships by Dzombak and Morel.[11] Hydrous ferric oxide, which is also known as amorphous ferric hydroxide, has a weak x-ray diffraction pattern similar to the natural occurring mineral ferrihydrite. The selectivity of hydrous ferric oxide for cations and anions is probably similar to the selectivity of ferrihydrite, which is a common weathering product of iron-containing minerals. The intrinsic cation adsorption constants for this solid show that several of the common trace metals (especially Cu^{2+} and Pb^{2+}, but also Ni^{2+} and Zn^{2+}) are much more strongly attracted (by orders or magnitude) to this surface than the normal major divalent cations in groundwater (Ca^{2+} and Mg^{2+}). Data from Schindler and Stumm[14] also show the strong affinity of other metal oxides ($\gamma\text{-}Al_2O_3$ and TiO_2) for the trace metals. Similarly, some of the trace anions ($Se_2O_3^{2-}$, MoO_4^{2-}, CrO_4^{2-}) appear to be more strongly attracted to the surface of hydrous ferric oxide than sulfate. These data show that the mobility of many of the trace elements in groundwater may be controlled by their strong specific adsorption onto surface complexation sites. The actual concentration of adsorbed species will be a function of many system variables (such as pH, Eh, solution species concentration, and concentrations of competing species); however, the surface complexation methodology provides a technique for accounting for all of these variables in calculating adsorption.

A wide variety of models have been developed to simulate surface complexation over the wide range of environmental conditions. These include the diffuse layer model,[15] generalized two-layer model,[11] constant-capacitance model,[16] and the triple-layer model.[17] Computer codes that contain surface complexation models along with equilibrium aqueous speciation models include MINTEQ,[18,19] SOILCHEM[20,21] and PHREEQC.[22] Mass transport models that contain surface complexation modeling capabilities include TRANQL[23] and HYDROGEOCHEM.[24] Geochemical modeling codes are discussed in Chapter 5 and Goldberg[25] provides a survey of adsorption models in chemical equilibrium computer codes.

Adsorption Final Remarks

This discussion of adsorption has covered the topics of ion exchange, adsorption isotherms, and surface complexation. The impact of cation exchange on a system can be reasonably well predicted from a knowledge of the cation exchange capacity and the concentrations of major cations. For clay minerals, the reaction is not particularly pH sensitive; however, if organic matter provides a significant portion of the exchange capacity, then pH will be a more important variable. The pH is also important in determining the available anion exchange capacity, which is generally much lower than the CEC.

Adsorption isotherms and the distribution coefficients derived from them can be useful snapshots of the affinity of a solid surface for trace solutes; however, they are very limited to the experimental conditions under which they are derived. Also, the method of determining the distribution coefficient does not allow for an understanding of the mechanism

TABLE 3-4.

EXPERIMENTALLY DERIVED AND ESTIMATED SURFACE COMPLEXATION CONSTANTS FOR CATIONS AND ANIONS ON HYDROUS FERRIC OXIDE

Cation	Log K_{int}	Anion	Log K_{int}
Ba^{2+}	−7.2	SO_4^{2-}	0.79
Sr^{2+}	−6.58	SeO_4^{2-}	0.80
Ca^{2+}	−5.85	SeO_3^{2-}	5.17
Mg^{2+}	−4.6	CrO_4^{2-}	3.9
Mn^{2+}	−3.5	MgO_4^{2-}	2.4
Co^{2+}	−3.01	F^-	1.6
Ni^{2+}	−2.5		
Cd^{2+}	−2.9		
Zn^{2+}	−1.99		
Cu^{2+}	0.6		
Pb^{2+}	0.3		

Note: Reactions: $\equiv FeOH + M^{2+} \leftrightarrow \equiv FeOM^+ + H^+$;

$\equiv FeOH + A^{2-}$ (or F^-) $\leftrightarrow \equiv FeOHA^{2-}$ or $\equiv FeOHF^-$.

Data from Dzombak and Morel.[11]

of adsorption so that predicting distribution coefficients for other environmental conditions is best considered an artistic pastime. The adsorption of dissolved hydrophobic organic compounds onto solid organic matter in the soil is not as sensitive to site conditions; therefore the K_{oc} values appear to be applicable to a wider range of site conditions.

To adequately estimate adsorption of trace elements for a range of site conditions, the surface complexation approach must be used. This method is valid over a wide range of pH conditions (as well as Eh and competing species conditions) because it includes the adjustment of surface conditions on the adsorbent solids for changes in the system. Also, if associated aqueous redox and complexation calculations are made, the appropriate species will compete for the available surface sites. The current hindrance to broader appeal of the surface complexation approach is the lack of a consistent set of intrinsic adsorption constants for the suite of solids present in the subsurface. A dataset has been developed by Dzombak and Morel[11] for hydrous ferric oxide, which is one of the most important solids for the adsorption of trace elements in the environment. It is expected that similar datasets will be developed in the future for other important adsorbents.

3.2. MINERAL PRECIPITATION/DISSOLUTION

The mineralogy of the subsurface and the general process of mineral alteration in response to chemical weathering were discussed in Section 1.3. The following section describes mineral equilibrium and solubility in the aquifer environment.

Mineral Equilibrium

A mineral in contact with groundwater represents a geochemical system consisting of a solid phase and a solution phase. If the solution does not initially contain any of the components of the mineral, then disequilibrium exists between the phases, and some of the mineral will dissolve to provide its components to the solution. As described in the discussion on chemical equilibrium (Section 1.2), the mineral may equilibrate with the water at which point dissolution and precipitation of the solid are in balance and solution concentrations of the mineral components are fixed at the equilibrium values. The barite/water system can be used to illustrate the equilibrium condition. Figure 3-9 shows the addition of the mineral barite ($BaSO_4$) to a beaker of water. The dissolution reaction for barite can be written as follows:

$$BaSO_4 \rightarrow Ba^{2+} + SO_4^{2-} \qquad (3\text{-}20)$$

(Note that none of the components of water are part of the reaction; therefore water does not need to be included in the reaction.)

Each barite molecule that dissolves from the solid provides an atom of barium and a molecule of sulfate to the water. The concentrations of barium and sulfate increase in solution with dissolution of barite until the solution concentration is saturated with the components of barite. At the equilibrium concentration values, barite may continue to dissolve but corresponding amounts of barium and sulfate will combine in solution to precipitate barite. As a consequence, the solution concentration does not change. This equilibrium can be represented as follows:

$$BaSO_4 \leftrightarrow Ba^{2+} + SO_4^{2-} \qquad K = 10^{-9.98} \left(25°C\right) \qquad (3\text{-}21)$$

where the arrow pointing in both directions signifies equilibrium between the solid and its dissolved components.

The equilibrium constant for the dissolution reaction can be written in terms of the activities of the dissolved components of the mineral.

$$K = \left(a_{Ba^{2+}}\right)\left(a_{SO_4^{2-}}\right) = 10^{-9.98} \qquad (3\text{-}22)$$

(The activity of the solid barite is considered equal to one.)

We can ignore aqueous complexation in this simple system because it turns out that the concentrations of the only significant ion pairs, HSO_4^- and $BaOH^+$, are orders of magnitude lower than the SO_4^{2-} and Ba^{2+} concentrations. The effect of ion shielding can also be ignored because the solution will be very dilute (ionic strength = 4.1×10^{-5}), and, consequently, the activity coefficients will be close to one. In this case the activities of the dissolved species are approximately equal to the concentrations, and the equilibrium

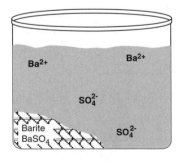

$$BaSO_4 \leftrightarrow Ba^{2+} + SO_4^{2-}$$

$$a_{Ba}{}^{2+}\, a_{SO_4}{}^{2-} = 10^{-9.98} \approx m_{Ba}{}^{2+}\, m_{SO4}{}^{2-}$$

Figure 3-9   Barite equilibrium in water.

constant expression can be used to approximate the total dissolved concentrations of barium ($m_{Ba^{2+}}$) and sulfate ($m_{SO_4^{2-}}$).

$$\left(m_{Ba^{2+}}\right)\left(m_{SO_4^{2-}}\right) \approx 10^{-9.98} \qquad (3\text{-}23)$$

If the system consists of only barite dissolving in water, then the only source of barium and sulfate to the solution is the dissolution of barite. Because barite dissolution provides an equal amount (on the mole scale) of barium and sulfate to solution, the dissolved concentrations of barium and sulfate will be equal in this system. Therefore,

$$m_{Ba^{2+}} = m_{SO_4^{2-}} \qquad (3\text{-}24)$$

Substituting this equality into Equation 3-23, produces

$$\left(m_{Ba^{2+}}\right)\left(m_{Ba^{2+}}\right) \approx 10^{-9.98} \qquad (3\text{-}25)$$

$$m_{Ba^{2+}} = 10^{-4.99} \text{ moles/L} \times 137 \text{ g/mole} \times 1000 \text{ mg/g} = 1.4 \text{ mg/L}$$

$$m_{SO_4^{2-}} = m_{Ba^{2+}} = 10^{-4.99} \text{ moles/L} \times 96 \text{ g/mole} \times 1000 \text{ mg/g} = 1 \text{ mg/L}$$

If the dissolved concentrations of barium and sulfate in our simple barite/water system each equal $10^{-4.99}$ moles per liter, then the solution concentration is saturated with the components of barite and the system is in equilibrium. If the concentrations of barium and sulfate in solution are less than $10^{-4.99}$ moles per liter, then the solution is undersaturated with respect to barite, and more of the solid will dissolve to attain equilibrium. If barium and sulfate were added to solution and concentrations achieved values higher than $10^{-4.99}$ moles per liter, then the solution would be oversaturated with respect to barite and it would be expected that the excess barium and sulfate would be removed from solution by precipitation of barite.

In the very simple barite-in-water system, the activities of dissolved barium and sulfate were equal; however, in most natural systems they will not be equal because there will be multiple sources of barium, sulfate and the other solutes. In this general case the solution concentration of the components of any mineral can be used to calculate whether the solution is in equilibrium with that mineral. This is done by comparing the ion activity product (IAP) of the mineral with the equilibrium constant of the mineral. The ion activity product for the barite dissolution reaction is simply the product of the activities of barium and sulfate in solution:

$$IAP_{barite} = \left(a_{Ba^{2+}}\right)\left(a_{SO_4^{2-}}\right) \qquad (3\text{-}26)$$

The ion activity product for any mineral is equivalent to the representation of the equilibrium constant for the reaction in terms of the activities of the reactants and products. Figure 3-10 shows the equations for the ion activity products of several common minerals. The ion activity product is calculated from solution concentration data for a water sample and then compared with the equilibrium constant for the mineral at the *in situ* water temperature. If the IAP is equal to the equilibrium constant, then the solution has the correct concentrations of the components of the mineral to be in equilibrium with the mineral. In the case of the simple barite/water system, the ion activity product at equilibrium is as follows:

$$IAP_{barite} = \left(a_{Ba^{2+}}\right)\left(a_{SO_4^{2-}}\right) = \left(10^{-4.99}\right)\left(10^{-4.99}\right) = 10^{-9.98} = K \qquad (3\text{-}27)$$

However, any combination of activities of dissolved barium and sulfate that produce an IAP of $10^{-9.98}$ will be in equilibrium with barite. If the barium activity is low (say 10^{-6} mol/L), then the sulfate concentration must be high ($10^{-3.98}$ mol/L), and vice versa, to maintain equilibrium with barite.

When the IAP for a mineral is less than its equilibrium constant ($K_{mineral}$), then the solution is undersaturated with respect to the mineral, and when the IAP is greater than $K_{mineral}$, then the solution is oversaturated with the components of the mineral. Calculation of a mineral saturation index (SI) is a convenient method of representing the equilibrium condition of a solution with respect to a mineral.

$$\text{Saturation index} = \log_{10} \frac{IAP}{K_{mineral}} \qquad (3\text{-}28)$$

SI = 0; mineral is in equilibrium with solution
SI < 0; mineral is undersaturated
SI > 0; mineral is oversaturated

Mineral equilibrium calculations for a groundwater sample are useful in predicting the presence of reactive minerals in the groundwater system and estimating mineral reactivity. If certain minerals such as calcite, ferrihydrite, and barite are commonly found in equilibrium with groundwater, then it is reasonable to assume that these minerals are reactive in typical groundwater environments and they can control solution concentration. A list of commonly found reactive minerals is given in Appendix A. If a groundwater sample is analyzed and the saturation index calculates to be near zero for one of the reactive minerals, then it is likely that the mineral occurs in the aquifer environment and is affecting solution composition. Because of uncertainties inherent in the calculation of saturation indices (such as the accuracy of the chemical analysis and mineral equilibrium constant and the method of calculating ion activities), a range of values for SI near zero is generally considered to be within the equilibrium zone for a mineral. Ranges of SI = 0 ± 0.5 and 0 ± (5%)(log $K_{mineral}$) have been used in various studies.[26]

By using the saturation index approach it is possible to predict the reactive mineralogy of the subsurface from the groundwater data without collecting samples of the solid phase and analyzing the mineralogy. Although validating a prediction from the groundwater data that a reactive mineral occurs in an aquifer should be done by analyzing the solid phase,

Calcite $CaCO_3 \longleftrightarrow Ca^{2+} + CO_3^{2-}$; IAP $= (a_{Ca^{2+}})(a_{CO_3^{2-}})$

Dolomite $CaMg(CO_3)_2 \longleftrightarrow Ca^{2+} + Mg^{2+} + 2CO_3^{2-}$; IAP $= (a_{Ca^{2+}})(a_{mg^{2+}})(a_{CO_3^{2-}})^2$

Pyrite $FeS_2 \longleftrightarrow Fe^{2+} + 2S^-$; IAP $= (a_{Fe^{2+}})(a_{S^-})^2$

Ferrihydrite $Fe(OH)_3 + 3H^+ \longleftrightarrow Fe^{3+} + 3H_2O$; IAP $= \dfrac{(a_{Fe^{3+}})}{(a_{H^+})^3}$

Gypsum $CaSO_4 \bullet 2H_2O \longleftrightarrow Ca^{2+} + SO_4^{2-} + 2H_2O$; IAP $= (a_{Ca^{2+}})(a_{SO_4^{2-}})$

Figure 3-10 Ion activity products for common minerals.

in many cases the reactive mineral cannot be identified in the solid because it is present at a concentration below the detection limit of the analytical method. The reactive mineral may be a weathering product that coats other grains or acts as a cement between grains, in which case it may form a very small percentage of the whole rock.

If the saturation index for a mineral calculates to be less than zero, the water is undersaturated with respect to that mineral. This means that the mineral cannot precipitate from solution, and should dissolve, if present, into solution to reach equilibrium concentrations. In an aquifer situation, where groundwater residence times are assumed to be on the scale of years to decades or longer, it is unlikely that a reactive mineral would remain undersaturated in the water. The most likely interpretation of an SI < 0 for a mineral is either that the mineral is not present if it is reactive or, if it is present, then it is not reactive. For example, calcite is generally considered to be a reactive mineral. If the SI for calcite for a groundwater sample shows the water to be undersaturated with calcite, then, assuming that the analytical data are accurate, it can usually be concluded that calcite is not present in the aquifer in contact with the groundwater. Equilibration of groundwater with a mineral is the primary criterion for deciding whether or not a mineral is reactive in the aquifer environment.

Using similar reasoning for the case of oversaturation, if the SI of a mineral is greater than zero, then the mineral is not reactive. If the mineral was reactive, it would limit solution concentration of its constituents to values that would produce an SI close to zero. The iron minerals ferrihydrite [$Fe(OH)_3$] and hematite (Fe_2O_3) provide a good example of this situation. Hematite is the more thermodynamically stable solid form of Fe(III), and, consequently, would limit dissolved iron to a lower concentration than ferrihydrite. However, oxidizing groundwaters are generally in equilibrium with ferrihydrite (SI $\approx$ 0), not hematite. In fact, saturation indices for hematite oftentimes are greater than 10, which represents oversaturation by a factor of 10^{10}. Hematite does not precipitate fast enough, even on the scale of groundwater residence time, to control the dissolved iron concentration; thus it is not a reactive mineral from the standpoint of limiting the dissolved iron concentration. Ferrihydrite may convert to the more stable hematite form of iron given sufficient time, but this process occurs by a mechanism that does not limit dissolved iron concentration to the hematite saturation value. The transformation/replacement of a more soluble and less stable precursor mineral such as ferrihydrite by a less soluble but more stable mineral such as hematite is known as Ostwald ripening.[27,28]

A word of caution concerning the use of saturation indices to identify reactive minerals is necessary at this point. It is possible that by some fortuitous set of circumstances that the concentrations in groundwater of the components of a mineral will just happen to be at the right combination of values to calculate a saturation index of close to zero even though the mineral is not forming in the aquifer environment. For example, diopside is a pyroxene mineral with a composition of $CaMgSi_2O_6$. Diopside forms in metamorphic rocks generally at temperatures greater than 600°C.[29] It is very unlikely that diopside would ever form at aquifer temperatures; however, it is possible that the calcium, magnesium, and silica concentrations in groundwater could be in the correct ratios, due to control of solution concentrations by other minerals, to calculate a saturation index for diopside close to zero. In this case diopside is not affecting the concentration of any of its components because it is not forming in the aquifer environment. It is just by chance that the concentrations of the components of diopside happen to be at particular values that calculate a saturation index of near zero for the mineral.

Data Display

Various methods of displaying and evaluating the mineral equilibrium condition of an aquifer are shown in Figure 3-11. In Figure 3-11a the saturation index of calcite has been plotted for several groundwater samples collected at a site versus the specific conductance of the water samples. The parameter chosen for the x-axis is designed to provide some spread to the data and to equate to some important variable (such as residence time) that might affect the solution concentration and, hence, mineral equilibrium. The saturation index zone representing calcite equilibrium is 0 ± 0.4 (5% of log $K_{calcite}$). The points plotted on the figure could represent locations along the flow path. Near the recharge area where specific conductances (dissolved concentrations) are low and the water is undersaturated with calcite, the groundwater may not have had sufficient time to equilibrate with calcite or, if calcite was originally present, it has been leached from the solid phase by the fresh recharge water. When specific conductance is higher, the majority of water samples appear to be in equilibrium with calcite, which suggests that calcite is present in the aquifer. There are two outliers shown on Figure 3-11a. The water sample that has an SI of 1 is suspect because it is unlikely that groundwater would be this far oversaturated with such a reactive mineral as calcite. An SI this high for calcite suggests that the analytical data and sample pH should be checked for errors, as well as the input dataset for the SI calculation. In instances such as these, the calculation of SI may identify an error in sampling, analyzing, or recording data. The water sample with the high specific conductance and low SI may be a correct value, but it is anomalous compared with adjacent values; therefore the data and calculations should also be rechecked. If the SI is shown to be correct, then other possibilities for the anomalously low SI value should be evaluated. It may be an indication of localized geochemical conditions due to natural or manmade causes and could identify a source of contamination.

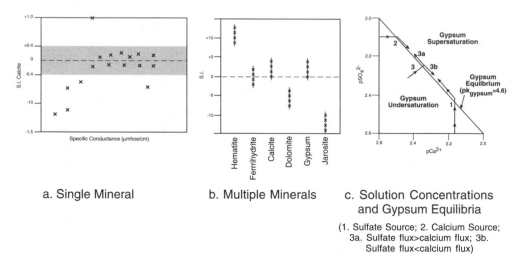

a. Single Mineral   b. Multiple Minerals   c. Solution Concentrations
                                                            and Gypsum Equilibria

(1. Sulfate Source; 2. Calcium Source;
3a. Sulfate flux>calcium flux; 3b.
Sulfate flux<calcium flux)

Figure 3-11 Saturation index presentations.

In Figure 3-11b the saturation indices of several minerals are plotted on a single diagram. Although this type of plot does not as easily allow for identification of trends at a site, it does provide a good method of displaying a considerable amount of data. Groundwater at this site appears to be in equilibrium with ferrihydrite, calcite, and gypsum. The data for ferrihydrite and hematite clearly show the separation between SIs and point to the likelihood that ferrihydrite is limiting the dissolved iron concentration and not hematite.

Figure 3-11c is a plot of the activities in solution of the components of the mineral gypsum ($CaSO_4·2H_2O$). The scales of the axes are pCa^{2+} and pSO_4^{2-}, which are equal to $-\log(a_{Ca^{2+}})$ and $-\log(a_{SO_4^{2-}})$, respectively. This is analogous to the pH scale, which is $-\log(a_{H^+})$. The line for pK_{gypsum} represents the mineral equilibrium condition where $(a_{Ca^{2+}})(a_{SO_4^{2-}}) = 10^{-4.6}$ or $(pCa^{2+})(pSO_4^{2-}) = 4.6$. Water samples that plot along this line have the appropriate activities of calcium and sulfate to be in equilibrium with gypsum. Water samples that plot to the left of the pK_{gypsum} line are undersaturated, and those that plot to the right are oversaturated with respect to gypsum. In addition to saturation information, this data presentation is also useful for evaluating chemical evolution of the groundwater. If water samples taken along the flow path follow the trajectory shown by line 1, then it would appear from the data that there is a source of sulfate to the water (for example, from oxidation of pyrite), but no additional source of calcium because the calcium activity stays constant until the groundwater data intersect the gypsum solubility line. When saturation with gypsum is reached, the mineral precipitates. If sulfate continues to be added to the water from its source, then the precipitation of gypsum lowers the activity of calcium and limits its solution concentration. Note that the precipitation of gypsum does not limit the activity of sulfate, which continues to increase along line 1. The opposite case of a source for calcium along the flow path (say from cation exchange) and no source for sulfate is shown by line 2. Line 3 shows a situation where there are sources of both calcium and sulfate to the groundwater along its flowpath. The activities increase until gypsum equilibrium is reached, at which point the evolutionary path may go up or down the slope of the equilibrium line. If sulfate is being supplied to the water at concentrations (on the mole scale) greater than calcium, then precipitation of gypsum will drive down the dissolved calcium activity while sulfate activity continues to increase (line 3a). If the flux of calcium exceeds the flux of sulfate to the water, then path 3b will be followed. This type of data presentation is useful in determining which component of a mineral is being limited by formation of the mineral.

Practical Calculation of Saturation Indices

Groundwater is a solution composed of numerous constituents that may be in equilibrium with a large variety of minerals. Because of the influence of all the dissolved species either directly (by complexation) or indirectly (by ion shielding) on the activities of the components of the minerals, a complete water analysis is required for accurate calculation of saturation indices. The general procedure for calculating mineral saturation indices for a water sample is illustrated in Figure 3-12. The process starts by collecting a water sample and measuring the dissolved concentration of its components. Components are the dissolved constituents (such as Ca^{2+}, Mg^{2+}, Na^+, CO_3^{2-}, SO_4^{2-}, etc.) from which all of the aqueous complexes can be calculated. The temperature and master variables pH and Eh must also be measured

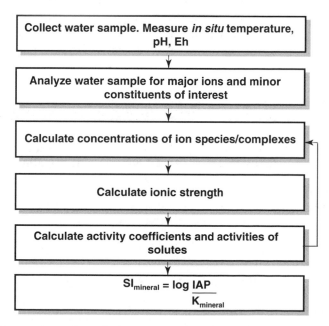

Figure 3-12 Mineral saturation index calculations.

for the water sample because they affect aqueous complexation reactions. The concentrations of ion species and complexes are calculated and subtracted from the total concentration of the component. The activity coefficient of the component is calculated from the solution ionic strength and then multiplied by the concentration of the uncomplexed component to give the activity of the free components. This activity is then used to calculate the ion activity products and saturation indices of possible minerals that might be in equilibrium with the solution. Because of the iterative nature of the calculations (the activity of a component is a function of the concentrations of its complexes, which are, in turn, calculated from the activity of the components of the complex), saturation indices are normally calculated using the computer codes described in Chapter 5.

In the discussion on aqueous complexation (Section 2.1), the concentrations of complexes found in a typical $Ca–HCO_3$ groundwater were calculated using the MINTEQA2 computer code. The output of the calculation also provides the saturation indices (not shown) for all of the minerals (59) in the database of the code capable of being formed from the solution composition. Because there are only 12 components in this water sample, there cannot be more than 12 minerals or other phases controlling the solution concentration level. In fact, 36 minerals have a calculated SI less than −1, and 12 minerals have an SI greater than 1. None of these 48 minerals are likely to be controlling solution concentration unless there are errors in the reported analyses.

Table 3-5 shows the saturation indices for some of the potential minerals that might be limiting solution concentration in the $Ca–HCO_3$ groundwater. The fact that both calcite and ferrihydrite $[Fe(OH)_3]$ have a calculated SI greater than one suggests a problem with the data. These are normally both reactive minerals and should be in equilibrium with the groundwater composition. The most likely explanation for at least part of the calculated oversaturation of these two minerals is that the reported pH is a laboratory measured value and is not representative of aquifer conditions. The calculated partial pressure of carbon dioxide gas in equilibrium with the groundwater is $10^{-3.0}$ atm, which is relatively close to the atmospheric value of $10^{-3.5}$ atm. As discussed previously (Section 2.3), typical P_{CO_2} values for aquifers are 10 to 100 times greater than atmospheric values. Table 3-5 shows the saturation indices and pH recalculated at P_{CO_2} values of $10^{-2.5}$, $10^{-2.0}$, and $10^{-1.5}$ atm. If the groundwater had been in equilibrium with a P_{CO_2} pressure of $10^{-2.0}$ atm, the pH of the water would be 7.3 and the saturation index for calcite would be 0.06.

If the actual aquifer P_{CO_2} pressure is $10^{-2.0}$ atm, the saturation index for ferrihydrite would be 2.2 instead of the original value of 2.3; therefore carbon dioxide degasing and pH increase cannot explain the high SI for this mineral. It is possible that the calculated

TABLE 3-5.

SATURATION INDICES FOR CA–HCO₃ GROUNDWATER				
Mineral	Original Data* (pH = 8.4)	P_{CO_2} set at 1E-2.5 atm	P_{CO_2} set at 1E-2.0 atm	P_{CO_2} set at 1E-1.5 atm
Calcite (CaCO₃)	1.1	0.55	0.06	−0.4
Dolomite [CaMg(CO₈)₂]	1.9	0.74	−0.24	−1.23
Ferrihydrite [Fe(OH)₈)]	2.3	2.4	2.2	1.7
Hematite (Fe₂O₈)	17.2	17.4	16.9	16.1
Chalcedony (SiO₂)	−0.08	−0.08	−0.08	−0.08
Christobalite (SiO₂)	0.02	0.02	0.02	0.02
SiO₂ (A, Glass)	−0.59	−0.59	−0.59	−0.59
Magnesite (MgCO₃)	0.29	−0.29	−0.8	−1.3
Calculated pH		7.8	7.3	6.8
Calculated P_{CO_2} (atm)	1E-03			

* Data from Snoeyink and Jenkins, *Water Chemistry*, John Wiley & Sons, New York, 1980.

oversaturation of ferrihydrite is a consequence of particulate iron being present in the sample water when it was collected and acidified for storage prior to analysis. Acidification to a pH of 2 would have dissolved some, if not all, of the suspended iron minerals in the sample, which would then show up as an erroneously high dissolved iron value for the sample. The water would appear to be oversaturated with respect to ferrihydrite as shown by the elevated SI. Note that the SI for hematite, the most stable form of iron, is greater than 16 for all P_{CO_2}/pH combinations, which shows that this mineral cannot be limiting the solution concentration for iron.

Other minerals shown by the calculation of SI to potentially affect solution concentration because they have values near zero are the silica solids (chalcedony, cristobalite, and amorphous silica glass) and magnesite ($MgCO_3$). Magnesite becomes increasingly undersaturated as the P_{CO_2} increases. Consequently it may not be present in the aquifer if the carbon dioxide partial pressure is closer to a value of 10^{-2} atm.

Mineral Solubility

The solubility of a mineral is the amount of the mineral that dissolves in water to establish equilibrium between the solid phase mineral and its components in solution. For the simple barite/water system it was found above that the equilibrium concentrations of barium and sulfate in solution were about $10^{-4.99}$ mol/L. Because the only source of barium and sulfate in this system is barite, this must also be the amount of barite that dissolved to reach equilibrium; therefore it is the solubility of barite in pure water. If the solution to which barite was added already contained barium or sulfate (or both), then the solubility of barite would be reduced because less would need to dissolve to reach equilibrium. For example, if gypsum ($CaSO_4{\cdot}2H_2O$) had equilibrated with the water before barite was added, the initial sulfate activity might be on the order of $10^{-2.3}$ mol/L. Using the relation between activity product and equilibrium constant for barite,

$$\left(a_{Ba^{2+}}\right)\left(a_{SO_4^{2-}}\right) = K_{barite} = 10^{-9.98} \tag{3-29}$$

the amount of barite that would dissolve to equilibrate with the gypsum-equilibrated solution would be approximately equal to the dissolved activity of barium at barite equilibrium:

$$a_{Ba^{2+}} = K_{barite}\Big/\left(a_{SO_4^{2-}}\right) = 10^{-9.98}\big/10^{-2.3} = 10^{-7.68} \text{ moles/L} = \text{barite solubility}$$

(This assumes that activities are equal to concentrations and the dissolution of barite does not significantly increase the dissolved sulfate concentration.)

In this case the presence of sulfate at $10^{-2.3}$ mol/L lowered the solubility of barite by over two orders of magnitude (from $10^{-4.99}$ to $10^{-7.68}$ mol/L). This is an example of the *common-ion effect* on solubility, where the sulfate ion common to both barite and gypsum lowers the solubility of barite. Barite solubility was decreased significantly because of the relatively high solubility of gypsum. If gypsum had been added to a solution in equilibrium with barite, the initial sulfate concentration ($\approx 10^{-4.99}$ mol/L) would depress the solubility of gypsum, but only by a relatively small amount because the sulfate equilibrium concentration for gypsum ($\approx 10^{-2.3}$ mol/L) is much greater than the initial sulfate concentration at barite equilibrium. In general, the impact of the *common-ion effect* is that the solubility of a mineral will be less if one of its components is present in solution when the solution contacts the mineral. The *common-ion effect* of calcium in calcite and gypsum on the solubilities of these minerals is discussed by Wigley.[36] Wigley also describes the probable importance of concurrent solution and precipitation of calcite and gypsum in karst terrain.

The solubility of a mineral may be a function of the temperature, pH, Eh, concentration of complexing ligands, and ionic strength of the solution. The equilibrium constants for the minerals are all temperature dependent. Most minerals are more soluble at higher temperatures; however, the carbonate minerals become less soluble as the temperature increases. Table 3-6 shows the effects of increasing temperature on simple systems of gypsum/water and calcite/water. Increasing the temperature from 10°C to 75°C increased the solubility of gypsum by a factor of 1.3 and decreased the solubility of calcite by a factor of 3.4. (All of the calculations of mineral solubility in this section were made with the MINTEQA2 code.[19])

TABLE 3-6.

EFFECT OF TEMPERATURE ON GYPSUM AND CALCITE SOLUBILITY (PH = 7.5)		
Temperature (Degrees C)	Gypsum solubility (mg/L)	Calcite solubility (mg/L)
10	1460	250
25	1570	190
50	1720	120
75	1920	76

The pH of the system affects solubility because it governs aqueous complexation. For minerals such as the carbonates that form several ion species (H_2CO_3, HCO_3^- and CO_3^{2-}) and strong complexes in the pH range of groundwater, pH can have a large impact on solubility. For other minerals such as the simple sulfate salts (for example, barite and gypsum) that do not form strong complexes with hydrogen and hyroxyl species at normal groundwater pH values, the pH does not significantly affect solubility. Table 3-7 provides the solubility of the calcite/water system closed to the exchange of CO_2 gas. Because the speciation of inorganic carbon is pH dependent and favors the carbonate ion at high pH, the solubility of calcite is depressed as pH increases. The solubility decreases by a factor of almost 50 as the pH increases from 6.5 to 9.5. The solubility data for gypsum listed in Table 3-7 show little effect due to pH because the calcium and sulfate complexes ($CaOH^+$ and HSO_4^-) formed in water are not present in high concentrations in the pH range of 6.5 to 9.5. At a pH of 9.5, the solubility of gypsum increases slightly because of the increase in concentration of the $CaOH^+$ complex.

Minerals commonly contain only one of the redox states of an element, for example, Fe (III) in ferrihydrite [$Fe(OH)_3$]. If the system is in redox equilibrium, its redox potential controls the concentration ratio of the redox states [for example, Fe(II)/Fe(III)] of the redox-

TABLE 3-7.

EFFECT OF PH ON CALCITE AND GYPSUM SOLUBILITY (TEMPERATURE = 25°C)		
pH	Calcite solubility (mg/L)	Gypsum solubility (mg/L)
6.5	890	1570
7.5	190	1570
8.5	54	1570
9.5	18	1689

TABLE 3-8.

EFFECT OF EH ON FERRIHYDRITE SOLUBILITY (PH = 7.5)	
Eh (mV)	Ferrihydrite solubility (mg/L)
300	0.0009
200	0.002
100	0.06
0	3.0

sensitive element. As a consequence the redox potential indirectly controls the solubility of minerals with redox-sensitive elements. Table 3-8 lists the solubility of ferrihydrite at a pH of 7.5 and several Eh values. Because the dominant iron redox state in solution changes from Fe(III) to Fe(II) over this range of Eh values, the solubility of the Fe(III) mineral increases by orders of magnitude (>3000x) as the redox potential decreases from +300 mV to 0 mV.

If the components of a mineral form complexes with other dissolved species in solution, then the effective dissolved concentrations (activities) of the components are reduced requiring more of the mineral to dissolve to achieve equilibrium concentration levels. Therefore the higher the concentration of complexing species present in groundwater, both inorganic and organic, the greater the solubility of minerals with components that form complexes. The divalent (Ca^{2+}, Mg^{2+}, SO_4^{2-}, CO_3^{2-}) and trivalent (Fe^{3+}, Al^{3+}, PO_4^{3-}) ions form fairly strong complexes with each other; therefore their presence in solution can increase the solubility of minerals containing these components. Table 3-9 shows the effect of magnesium in water on the solubility of barite. The formation of the $MgSO_4^0$ complex in solution increases the solubility of barite by a factor of 5 over a range of magnesium concentrations from 0 to 0.1 mol/L (2430 mg/L).

TABLE 3-9.

EFFECT OF COMPLEXATION ON BARITE SOLUBILITY		
Mg^{2+} [mol/L (mg/L)]	Barite solubility (mg/L)	$MgSO_4^0$ (mol/L)
0	2.5	0
0.001 (24.3)	3.1	1.4×10^{-6}
0.01 (243)	5.3	8.3×10^{-6}
0.1 (2430)	12.8	3.4×10^{-5}

The ionic strength of a solution is a measure of the ion shielding that occurs around charged, dissolved species. Ion shielding lowers the activity of dissolved species; therefore the higher the ionic strength, the greater the shielding and the greater the solubility of minerals in contact with the solution. Table 3-10 shows the effect of increasing ionic strength on the solubility of barite. The solubility increases by a factor of about 5 if the ionic strength increases from 4.2×10^{-5} to 1. For reference, the ionic strength of most dilute groundwater is in the range of 10^{-2} to 10^{-3}, while seawater has an ionic strength of 0.7.

TABLE 3-10.

EFFECT OF IONIC STRENGTH ON BARITE SOLUBILITY	
Ionic Strength (NaCl solution)	Barite solubility (mg/L)
4.2×10^{-5}	2.5
0.01	3.7
0.1	7.4
1	13

3.3. REACTION RATES

The use of equilibrium constants and distribution coefficients to calculate solution concentration as a function of mineral solubility, adsorption/desorption, and ion speciation/complexation assumes that the system comes to equilibrium with respect to all of its phases (solution, gas, and solids). Equilibrium is generally a good assumption for reactions that take place only in solution, such as aqueous complexation. The one notable exception is redox equilibrium, where it has been found that equilibrium does not commonly exist between the various redox couples and the measured redox potential (see Section 2.2). Adsorption of solutes onto the surfaces of solid phases is also generally considered a fast reaction, especially on the time scale of groundwater residence time. However, desorption has not been observed to be as rapid a process as adsorption,[30] except for the case of cation exchange reactions; therefore it is more difficult to achieve equilibrium concentrations in solution for most of the types of adsorption/desorption reactions.

The achievement of complete mineral equilibrium in groundwater systems between solution concentrations and all the minerals in contact with the groundwater is rarely achieved. As discussed previously (see Chemical Weathering, Section 1.3), the primary minerals in many rock types will never equilibrate with water under aquifer pressure and temperature conditions because they only form at very high temperatures and, perhaps, pressures. These minerals dissolve by chemical weathering processes, thereby releasing their constituents to the water. Secondary minerals may become oversaturated as the primary minerals dissolve, resulting in the precipitation of new minerals. The rate of precipitation of these secondary minerals will determine whether they can control solution concentrations.

Because of the disequilibrium generally found in aquifer systems between groundwater and many of the minerals comprising the aquifer and between the redox-sensitive elements found in solution and in some of the minerals, it is necessary, when using the equilibrium assumption to predict solution concentrations, to choose only those minerals and redox pairs that are reactive and can equilibrate with the system. For example, because olivine and pyroxene minerals do not form in an aquifer environment, it is not realistic to assume that they will control groundwater concentration in an aquifer even though they might be the dominant primary minerals. Likewise, the redox potential cannot usually be calculated from the dissolved oxygen concentration. Reactive minerals that may control groundwater composition were discussed in Section 1.3, and reactive redox pairs were discussed in Section 2.2.

The formation of a new mineral from solution requires that the mineral become initially oversaturated to some extent to produce the first solid surface of the mineral in the system. The amount of energy required to produce this initial surface is termed the activation energy of crystal formation. The higher the activation energy, the greater the amount of oversaturation required before crystallization can commence. The presence of other mineral surfaces in the aquifer may provide a template for new mineral formation. These surfaces serve as catalysts for the reaction by increasing the reaction rate without participating in the overall reaction stoichiometry. Enzymes in microorganisms are another important type of catalyst in aquifer systems. They are specialized proteins that catalyze specific biochemical reactions and enhance biodegradation of organic contaminants. Without biochemical catalysts, the biodegradation rates of most organic compounds would be much slower, and these contaminants would be much more persistent in the environment. For example, under sterile, oxidizing conditions benzene, toluene, and xylene can be stable in solution over a time period of months; however, when aerobic organisms are present these compounds were found to be almost completely degraded in about two months.[31]

Once a new mineral has begun to form or if seed crystals of that mineral are already present in the system, the rate of mineral formation is usually a function of the degree of disequilibrium between the solution concentration of the components of the mineral and the final equilibrium concentration. The higher the solution concentration compared with the equilibrium concentration, the greater the rate of mineral formation. As the equilibrium concentration is approached in solution, the rate of mineral formation slows down. This can be shown by the following equation[32]:

$$dC_t/dt = -k(C_t - C_{eq})^x \qquad (3\text{-}30)$$

This equation shows that the rate of precipitation (dC_t/dt) at any time t is a function of the departure of the current concentration (C_t) of a component of the mineral from the solution concentration (C_{eq}) at equilibrium with the mineral. The greater the excess concentration in solution, the faster the rate. The dependency on ($C_t - C_{eq}$) is commonly first or second order. If x = 1 (first order), the rate constant k has units of s^{-1}, and if x = 2 the units of k are $(mol/L)^{-1}s^{-1}$. The complementary rate expression for a dissolution reaction is as follows:

$$dC_t/dt = -k(C_{eq} - C_t)^x \qquad (3\text{-}31)$$

In this case the rate of dissolution of the mineral is enhanced by low, initial solution concentrations of the components of the mineral and high mineral solubility.

Other factors that may affect the rate of reaction are temperature, pH, and Eh. In general, most chemical reactions usually double or treble in rate for each rise in temperature of 10°C.[33] The increase in temperature facilitates movement and reaction of species dissolved in solution and movement of dissolved species to and from the surfaces of the solids. Also affecting the rate of reaction in response to temperature change is the change in mineral solubility and, perhaps, degree of disequilibrium produced by the change in temperature.

Laboratory and field data show that the rate of Cr(VI) reduction to Cr(III) by Fe(II) is inversely related to the pH. Cr(III) is the more stable redox species of chromium in most soil and groundwater environments, but Cr(VI) has been found to persist at higher concentrations than Cr(III) at contaminated sites because of the slow reduction rate of Cr(VI) at neutral and higher pH values. The half-life of Cr(VI) has been measured in the laboratory at a pH of about 5.5 to be on the order of a few months.[34] (Half-life is the amount of time necessary for half of the initial amount of Cr(VI) in solution to be reduced to Cr(III).) At a Cr(VI) contaminated site in Texas where the aquifer pH is 7, the half-life of Cr(VI) was calculated to be 2.5 years.[35] Accurate information on reaction rates can be an important factor in predicting the fate of contaminants in the environment.

REFERENCES

1. Brady, N.C., *The Nature and Properties of Soils,* Macmillan, New York, 1974.
2. Appelo, C.A.J. and Postma, D., *Geochemistry, Groundwater and Pollution,* A.A. Balkema, Rotterdam, 1994.
3. Sposito, G., *The Chemistry of Soils,* Oxford University Press, New York, 1989.
4. Magaritz, M. and Luzier, J.E., Water-rock interactions and seawater-freshwater mixing effects in the coastal dune aquifer, Coos Bay, Oregon, *Geochem. Cosmochim. Acta,* 49, 2515–2525, 1985.
5. Xue, Y. et al., Sea-water intrusion in the coastal area of Laizhou Bay, China. I. Distribution of sea-water intrusion and its hydrochemical characteristics, *Ground Water,* 31, 532–537, 1993.
6. U.S. Environmental Protection Agency, Soil Screening Guidelines: Technical Background Document. EPA/540/R95/128, Washington, DC, 1996.
7. Verschueren, K., *Handbook of environmental data on organic chemicals,* Van Nostrand Rheinhold, 1983.
8. Lyman, W.J., Reehl, W.F., and Rosenblatt, D.H., *Handbook of Chemical Property Estimation Methods,* American Chemical Society, Washington, DC, 1990.
9. Fetter, C.W., *Contaminant Hydrogeology,* Macmillan, New York, 1993.
10. Freeze, R.A. and Cherry, J.A., *Groundwater,* Prentice-Hall, Englewood Cliffs, NJ, 1979.
11. Dzombak, D.A. and Morel, F.M.M., *Surface Complexation Modeling. Hydrous Ferric Oxide,* John Wiley, New York, 1990, pp. 1–393.
12. Morel, F.M.M., *Principles of Aquatic Chemistry,* John Wiley, New York, 1983.
13. Bolt, G.H. and Riemsdijk, W.H.v., in *Soil Chemistry,* Bolt, G.H., Eds., Elsevier, Amsterdam, 1982, pp. 459–504.
14. Schindler, P.W. and Stumm, W., in *Aquatic Surface Chemistry,* Stumm, W., Eds., Wiley-Interscience, New York, 1987, pp. 83–110.
15. Stumm, W., Huang, C.P., and Jenkins, S.R., Specific chemical interactions affecting the stability of dispersed systems, *Croat. Chem. Acta,* 48, 223–245, 1970.
16. Schindler, P.W. and Gamsjager, H., Acid-base reactions of the TiO_2, anatase)–water interface and the point of zero charge of TiO_2 suspensions, *Kooloid Z.Z. Pollymere,* 250, 759–763, 1972.
17. Davis, J.A., James, R.O., and Leckie, J.O., Surface ionization and complexation at the oxide/water interface. I. Computation of electrical double layer properties in simple electrolytes, *J. Colloid Interface Sci.,* 63, 480–499, 1978.
18. Felmy, A.R., Girvin, D.C., and Jenne, E.A., *MINTEQ: A Computer Program for Calculating Geochemical Equilibria,* U.S. Environmental Protection Agency, 1983.
19. Allison, J.D., Brown, D.S., and Novo-Gradac, K.J., *MINTEQA2/PROEDFA2, a Geochemical Assessment Model for Environmental Systems,* U.S. Environmental Protection Agency, 1991.
20. Sposito, G. and Coves, J., *SOILCHEM: A Computer Program for the Calculation of Chemical Speciation in Soils,* Kearney Foundation of Soil Science, University of California, Riverside, 1988.
21. Sposito, G. and Coves, J., in *Chemical Equilibrium and Reaction Models,* Loeppert, R.H., Schwab, A.P., and Goldberg, S., Eds., Soil Science Society of America, Madison, WI, 1995, pp. 271–287.
22. Parkhurst, D.L., *User's Guide to PHREEQC, a Computer Model for Speciation, Reaction Path, Advective Transport and Inverse Geochemical Calculations,* U.S. Geological Survey, 1995.
23. Cederberg, G.A., Street, R.L., and Leckie, J.O., A groundwater mass transport and equilibrium chemistry model for multicomponent systems, *Water Resour. Res.,* 21, 1095–1104, 1985.
24. Yeh, G.-T. and Tripathi, V.S., *HYDROGEOCHEM: A Coupled Model of HYDROlogic Transport and GEOCHEMical Equilibria in Reactive Multicomponent Systems,* Oak Ridge National Laboratory, Oak Ridge, TN, 1990.
25. Goldberg, S., in *Chemical Equilibrium and Reaction Models,* Loeppert, R.H., Schwab, A.P., and Goldberg, S., Eds., Soil Science Society of America, Madison, WI, 1995, pp. 75–96.
26. Paces, T., *Kinetics of Natural Water Systems,* I.A.E.A., Vienna, 1975.
27. Morse, J.W. and Casey, W.H., Ostwald processes and mineral paragenesis in sediments, *Am. J. Sci.,* 288, 537–560, 1988.
28. Steefel, C.I. and Cappellen, P.V., A new kinetic approach to modeling water-rock interaction: The role of nucleation, precursors, and Ostwald ripening, *Geochim. Cosmochim. Acta,* 54, 2657–2677, 1990.

29. Press, F. and Siever, R., *Earth,* Freeman, W.H., San Francisco, 1974.
30. Jenne, E.A., in *Trace Inorganics in Water,* Gould, R.F., Eds., American Chemical Society, Washington, DC, 1968, pp. 337–387.
31. Barker, J.F., Patrick, G.C., and Major, D., Natural attenuation of aromatic hydrocarbons in a shallow sand aquifer, *Ground Water Monit. Remed.,* 64–71, 1987.
32. Nancollas, G.H., The growth of crystals in solution, *Adv. Coll. Interface Sci.,* 10, 215–252, 1979.
33. Krauskopf, K.B., *Introduction to Geochemistry,* McGraw-Hill, New York, 1979.
34. Rai, D. et al., *Chromium Reactions in Geologic Material,* Electric Power Research Institute, 1988.
35. Henderson, T., Geochemical reduction of hexavalent chromium in the Trinity Sand Aquifer, *Ground Water,* 32, 477–486, 1994.
36. Wigley, T.M.L., Chemical Evolution of the System Calcite-Gysum-Water, Can. J. Earth Sci., 10, 306–315, 1973.

4 DEVELOPMENT OF CONCEPTUAL GEOCHEMICAL MODELS

An understanding of the fundamental geochemical processes that influence the composition of an aquifer can be put to use by developing a conceptual model of the system. A conceptual geochemical model is the investigator's attempt to explain the chemical characteristics of the system (e.g., composition, pH, Eh) in terms of the water/rock/gas interactions active in the system. For example, for a limestone aquifer a conceptual model that involves calcite and dolomite dissolution/precipitation and CO_2 gas exchange may be developed to simulate the interaction of groundwater with soil gases and aquifer solids along the flow path and to explain the distribution and trends in pH, calcium, magnesium, and inorganic carbon concentrations. This chapter describes the data requirements and methods for developing conceptual geochemical models. The computer codes used to make the calculations required by the conceptual model are discussed in Chapter 5, and models of natural systems are described in Chapter 6.

4.1. MODELING DATA REQUIREMENTS

The geochemical model of a system attempts to simulate the chemical reactions in the system that produce the observed distribution of data. As a consequence, sufficient data must be available from the site to make the necessary calculations of chemical processes, and the model must include the equilibrium constants or rate constants for all the important reactions. For example, if gypsum is a reactive mineral in an aquifer, it is important to know the solubility of gypsum under the environmental conditions of the aquifer. Because sulfate forms relatively strong ion complexes with magnesium, which will increase the solubility of gypsum, it is necessary to measure the concentration of magnesium, and other major cations and anions, in the groundwater to accurately model gypsum in this system. It is also necessary to have a calculation method that incorporates the formation of the magnesium sulfate complex and other important complexes (e.g., $CaSO_4^0$ and $MgCO_3^0$) that compete for the dissolved magnesium and sulfate. Data requirements generally fall into two classes, site data on water/rock composition and experimentally derived data on equilibrium and rate constants.

Site Data Collection for Geochemical Modeling

The amount and type of site data required for developing a geochemical model depends, to some degree, on the purpose for studying the site. If the main focus of the study is to understand the distribution of the major ions in groundwater, then measuring their dissolved concentrations and the pH along the flow path of each identified aquifer may be adequate. The pH is necessary because of its importance on the solubility of many minerals. The Eh is probably not important because the typical major cations and anions (except for sulfate) in groundwater are not redox sensitive. If the aquifer contains zones that are potentially very reducing, then Eh may be an important measurement because of the possibility of reduction of sulfate to sulfide under reducing conditions. The measurement of minor and trace elements in the groundwater may or may not be necessary. If it is suspected that clay minerals are forming in the system and incorporating the major cations in their structure, then it is necessary to measure the minor/trace elements aluminum and silicon in solution so that the saturation indices of the clay minerals can be calculated. Also, if clay minerals are present in the aquifer, they may play an important role in major ion concentrations through cation exchange reactions. In this case the cation exchange capacity of the aquifer solids should be measured. This example shows that the selection of data to be collected in the field requires an understanding of the potential geochemical reactions that might impact water and rock composition for the elements of interest. Because the reactions are so interwoven and site specific, it is not possible to give a list of data requirements that covers all contingencies; however, general guidelines can be provided.

Table 4-1 provides a list of general groundwater data requirements for the development of a geochemical model of a system. Major cations and anions must be measured in all cases because of the formation of complex species between these ions and minor/trace dissolved species. Also, the concentrations of the major ions are used to calculate the ionic strength of the solution, which is a measure of ion shielding that lowers the activity of all dissolved ions. The pH of the groundwater is a necessary parameter because of its role in aqueous complexation and its affect on mineral solubility. If redox-sensitive elements are to be considered in the model, then the redox potential should be measured. The concentration of dissolved oxygen can be useful for evaluating the oxidizing capacity of the groundwater and is a good qualitative indicator of redox potential. The dissolved concentrations of silicon and aluminum must be known to calculate clay mineral equilibrium, and the concentrations of iron and manganese are necessary for calculating equilibrium with the oxyhydroxide minerals of these metals. Data on the other dissolved trace elements in

TABLE 4-1.

SITE GROUNDWATER DATA FOR GEOCHEMICAL MODELING

Data	Use
Major Ions	Calculation of solution complexes
(Ca, Mg, Na, K,	Calculation of ionic strength and solute activity
$HCO_3/CO_3/Cl/NO_3$)	Saturation indices for minerals with these components
pH	Ion speciation/complexation and mineral solubility
Eh	Ion speciation/complexation and mineral solubility of redox-sensitive elements
Dissolve Gases	
(O_2, CO_2)	O_2: qualitative measure of redox potential
	CO_2: stability of groundwater pH in contact with atmos. air
Minor/Trace Elements	
(Si, Fe, Mn, Al)	Clay and oxyhydroxide mineral equilibria
Trace Metals	
(Ba, V, Cr, Mo, Pb, Cu, Zn, Hg, Cd, B, etc.)	Mineral equilibria, competitive adsorption
Trace Semi-Metals	
(As, Se)	Mineral equilibria, competitive adsorption
Trace Non-Metals	
(F, Br, I, P)	Complexation, mineral equilibria, competitive adsorption
Organic Compounds	
(Humic/fulvic acids, etc.)	Complexation, oxygen consumption, sorption reactions
Stable Isotopes	
($^{18}O/^{16}O$, D/H $^{34}S/^{32}S$)	Water signature, mineral reactions
Unstable Isotopes	
Tritium, C-14, Cl-36	Age dating

groundwater should be collected if reactive minerals containing the element may be present (e.g., barium in barite), the element forms strong complexes (e.g., fluoride and phosphate), or the element competes for adsorption sites (e.g., As and Se). Naturally occurring organic compounds in groundwater, such as fulvic and humic acids, can complex with inorganic elements thereby impacting mineral solubility and the mobility of the species. Organic compounds in groundwater, both dissolved and particulate, also provide a strong reductant that can consume dissolved oxygen and lower the redox state of other dissolved species. Stable and unstable isotopes of the elements can be used to date groundwater and also discern some of the physical and chemical processes that might affect groundwater composition. For example, evapotranspiration changes the $\delta^{18}O$ and δD values from that consistent with the meteoric water line,[1] and $\delta^{13}C$ can be used to validate mass transfer predictions between groundwater, carbonate minerals, and CO_2 gas.[2]

Although much can be surmised about the aquifer solid phase from a complete analysis of the groundwater, it is recommended that solids predicted to be present from saturation index calculations be confirmed by analysis of the solid phase. Also, certain properties of the solids such as adsorption and neutralization capacities cannot be predicted from the solution analysis. Table 4-2 provides a list of the more common reactive minerals and characteristics of the solid phase that are potentially useful in developing a geochemical model of a system. Other reactive minerals that may be important in particular systems are listed in Appendix A. In addition to general composition, other features of the aquifer solid material that may be useful in developing a conceptual geochemical model and should be identified include (1) variations in composition and concentration along the flow path, (2) secondary minerals forming in the system in the void spaces of porous or fractured rock, (3) minerals that are being dissolved or leached by the groundwater, (4) exchangeable cations on the clays minerals and trace elements adsorbed onto metal oxyhydroxides and organic compounds, and (5) isotopic composition of the minerals and variations along the flow path.

Equilibrium and Reaction Rate Considerations for Geochemical Modeling

Site-specific groundwater data for the site to be modeled can be used to make speciation and mineral equilibrium calculations that will identify those solids in equilibrium with the solution that may be controlling the solution composition. To make these calculations, thermodynamic data (equilibrium constants and enthalpy values for the reactions) must be

TABLE 4-2.

SITE SOLID PHASE DATA FOR GEOCHEMICAL MODELING	
Constituent of Solid Phase	**Potential Impact on System**
Calcite	Mineral solubility control on solution concentration, partial measure of neutralization capacity
Gypsum	Mineral solubility control on solution concentration
Dolomite	Source of constiutents to solution, partial measure of neutralization capacity
Clay mineral identification, concentration & exchange capacity	Exchange sites for major cations, mineral solubility control on solution concentration
Ferric and manganese oxyhydroxide	Mineral solubility control on solution concentration, adsorption substrates for minor/trace elements
Pyrite	Mineral solubility control on solution concentration, source of acidity under oxidizing condition
Silicate minerals	Sources of many dissolved constituent
Organic carbon	Adsorbent medium for organic and inorganic compounds, reducing agent, source of dissolved carbon

available for all the potentially important chemical reactions. The types of reactions that are commonly important in groundwater systems are shown in Table 4-3. Because of the complexity of making these calculations for a multicomponent, multiphase system such as an aquifer, computer codes have been developed to speed the process. A thermodynamic database containing the necessary equilibrium constants and enthalpy values usually is associated with the computer codes. The databases typically cover all of the types of reactions listed in Table 4-3 except for solid solution processes, which are not commonly modeled. Representative computer codes available for geochemical modeling are described in Chapter 5.

TABLE 4-3.

COMMON WATER/ROCK/GAS INTERACTIONS

Process	Example Chemical Reaction
Gas equilibrium	$2CO_2(g) + H_2O \leftrightarrow CO_2(aq) + H_2CO_3$
Ion speciation	$H_2CO_3 \leftrightarrow HCO_3^- + H^+$
Ion complexation	$Ca^{2+} + HCO_3^- \leftrightarrow CaHCO_3^+$
Mineral dissolution/precipitation	$CaCO_3 \leftrightarrow Ca^{2+} + CO_4^{2-}$
Oxidation/reduction	$CH_2O + O_2 \leftrightarrow CO_2 + H_2O$
Adsorption/desorption	$\equiv FeOH_2^+ + Ag^+ \leftrightarrow \equiv FeOHAg^+ + H^+$
Solid solution	$CaCO_3 + xMg^{2+} \leftrightarrow Ca_{1-x}Mg_xCO_3 + xCa^{2+}$

Nonequilibrium conditions such as the slow dissolution or precipitation of a mineral relative to groundwater residence time require algorithms in the codes that incorporate rate constants for the reaction. None of the readily available geochemical modeling codes include the ability to explicitly consider reaction rates. However, in developing a conceptual model for a site that includes mineral equilibrium there is the implicit assumption that the chosen minerals are reactive and precipitation kinetics are fast enough that equilibrium will be achieved if the solution becomes saturated with the components of the mineral. For the case of dissolution kinetics, the codes allow a mineral to dissolve to equilibrium if the mineral is undersaturated in solution or they allow only a given amount of a mineral to dissolve in a unit mass of solution. The amount of the solid allowed to dissolve may or may not allow the solution to be saturated with the mineral. By choosing reactive minerals for the model and by having the opportunity to limit the amount that dissolves to correspond to the amount available in the aquifer, the modeler has the ability to simulate many natural systems where kinetics do not have a major impact on solution concentration. The assumption of partial equilibrium, where the groundwater is in equilibrium with some of the minerals present in the aquifer, is justified in natural groundwater systems where long residence times allow equilibrium to be achieved with reactive phases.

For some components in natural systems the rate of a reaction is a major factor in controlling solution composition. In the chromium example given in Section 11.4, the thermodynamically stable species is Cr(III) in Cr(OH)$_3$; however, the slow kinetics of chromium reduction from Cr(VI) to Cr(III) allow for disequilibrium to prevail for periods of years in an aquifer at neutral or higher pH.[3] Redox reactions for both organic and inorganic compounds are commonly slow, even on the groundwater residence time frame of decades or longer, unless a catalyst, often bacterial, is present to facilitate the process.

Other interactions such as desorption may also be relatively slow. In these cases it is useful to initially assume equilibrium conditions to establish expected concentrations at equilibrium. The calculated equilibrium concentrations can then be compared with field conditions to determine which parts of the system are not in equilibrium. Equilibrium calculations show the direction that the system will progress and how much additional reaction progress must occur before equilibrium is attained. Rate constants for these reactions may have to be developed from site data or from laboratory experiments. They should be compared with other reported values to confirm that the rates are reasonable.

4.2. ELEMENTS OF A CONCEPTUAL MODEL

A geochemical model of a system is a representation of the system in terms of the chemical reactions that occur between the various phases in the system. The elements of the model are the interacting phases and the amount of the phase that dissolves into or precipitates or exsolves from the solution. The phases are defined either by their initial and final composition for the solution phase(s) or by their fixed composition for the solid and gas phases. More than one solution phase may be considered if mixing of water types occurs along the flow path.

The aquifer shown in Figure 4-1 will be used to illustrate the development of a conceptual model. The aquifer occurs in a fractured marine claystone that has been exposed to near-surface weathering conditions since uplift of this portion of the California Coast Range over one million years ago. The major minerals in the claystone are opal, clays, feldspar, and quartz. The minor minerals are pyrite, calcite, and siderite in the unweathered portion of the rock and ferrihydrite, gypsum, and calcite in the weathered portion. The presence of alteration products (ferrihydrite and gypsum) in the weathered rock is a good indication that this is a reactive system. The reactivity of the marine claystone is partially due to the presence of ferrous iron minerals (pyrite and siderite) that formed in the sediment on the seafloor, but which are not stable under earth surface conditions. Oxygen in the air and dissolved in infiltrating water reacts with the ferrous iron minerals to form minerals stable under oxidizing conditions. The oxidation reactions for pyrite (FeS_2) and siderite ($FeCO_3$) can be written as follows:

$$2FeS_2 + 7.5O_2 + 7H_2O = 2Fe(OH)_3 + 4SO_4^{2-} + 8H^+ \qquad (4\text{-}1)$$

$$4FeCO_3 + O_2 + 10H_2O = 4Fe(OH)_3 + 4HCO_3^- + 4H^+ \qquad (4\text{-}2)$$

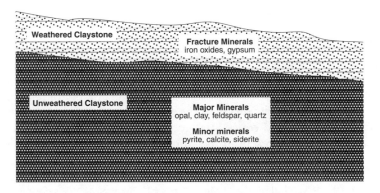

Figure 4-1   Mineralogy of weathered marine claystone.

These reactions can account for the occurrence of oxidized iron minerals [ferrihydrite, $Fe(OH)_3$] in the weathered zone, but do not account for the gypsum.

The oxidation of ferrous iron and sulfide sulfur would also produce an acidic solution unless other reactions are occurring that utilize the free hydrogen produced by these reactions. The free hydrogen is partially neutralized by calcite according to the following reaction:

$$CaCO_3 + H^+ \rightarrow Ca^{2+} + HCO_3^-$$ (4-3)

The dissolution of calcite releases calcium to solution, and the formation of the bicarbonate species reduces the activity of free hydrogen, thereby moderating the effect of pyrite and siderite oxidation on the pH of the system. As long as there is sufficient calcite present in the system to neutralize all of the hydrogen released by oxidation of pyrite and siderite, the solution will not become very acidic. The release of calcium by calcite dissolution and sulfate from oxidation of pyrite will increase the concentrations of these two components in solution until gypsum becomes saturated. At this point, additional dissolution of pyrite and calcite will result in the precipitation of gypsum to maintain calcium and sulfate concentrations at the equilibrium values. This reaction can be written as follows:

$$Ca^{2+} + SO_4^{2-} + 2H_2O \rightarrow CaSO_4 \cdot 2H_2O$$ (4-4)

It is useful in developing a conceptual geochemical model to consider the reactions that drive other chemical responses. These are generally the irreversible reactions that would consume the mineral given sufficient time and reactants. The minerals susceptible to irreversible reactions are not stable in the current environment. In the case of weathering of the marine claystone, pyrite, and siderite oxidation are irreversible reactions. They formed and were stable under the reducing conditions found in the clays accumulating with organic matter on the seafloor. Under earth surface conditions they will continue to dissolve as long as oxygen is available because they are unstable in an oxidizing environment. The reactions are not reversible because pyrite and siderite cannot form in an oxygenated environment. The oxidation of these minerals enhances the solubility of calcite and produces calcium and sulfate concentrations high enough to precipitate gypsum.

The occurrence of gypsum and oxidized iron minerals in the fractures of the weathered zone provides evidence that the conceptual reaction model reasonably explains the observed mineralogy. The oxidizing reactions continue to occur at the base of the weathered zone where water with dissolved oxygen comes into contact with pyrite and siderite in the claystone. The thickness of the weathered zone is a function of the rate at which oxygen is delivered to this boundary, the reducing capacity of the system, and the duration of oxidation.

The conceptual geochemical model of pyrite/siderite oxidation/dissolution followed by calcite dissolution and gypsum precipitation applies primarily to reactions at the interface of the weathered/unweathered claystone. Groundwater present in the weathered zone is not as influenced by oxidation/reduction reactions because of the small amount of remaining oxidizable minerals, but it is affected by reactions with the weathering products left behind by the earlier reactions. Therefore a second conceptual geochemical model is necessary to explain the current geochemical system present in the upper portion of the aquifer. Table 4-4 shows the major ion concentrations for two wells along the groundwater flow path. The upgradient well is near the recharge area and the downgradient well is approximately 1.5 km (1 mile) from the upgradient well. The Ca^{2+}, Mg^{2+}, Na^+, Cl^-, and SO_4^{2-} concentrations all increase significantly over this flow path, while HCO_3^- decreases and iron concentration and pH do not change appreciably. The conceptual model for this system must include

TABLE 4-4.

CLAYSTONE AQUIFER GROUNDWATER COMPOSITION (mg/L)

Component	Upgradient well	Downgradient well
Ca^{2+}	163	659
Mg^{2+}	89	448
Na^+	197	1370
Cl^-	315	2550
SO_4^{2-}	460	2430
HCO_3^-	353	280
Fe	<0.03	0.61
pH (std units)	6.9	6.7

sources for the constituents that increase in concentration and sinks for the constituents that decrease. The sinks must be minerals that precipitate fast enough to limit solution concentrations. The reactive minerals known to occur in this system are calcite, gypsum, and ferrihydrite. In addition to the calcite and gypsum equilibrium reactions given in Equations 4-3 and 4-4, the ferrihydrite reaction may limit the dissolved iron concentration:

$$Fe(OH)_3 + 3H^+ = Fe^{3+} + H_2O \qquad (4\text{-}5)$$

Clay minerals will participate in cation exchange by the following possible reactions, where X is a cation exchange site:

$$NaX + 1/2\,Ca^{2+} = Ca_{1/2}X + Na^+$$

$$NaX + 1/2\,Mg^{2+} = Mg_{1/2}X + Na^+$$

Gypsum and the oxidized iron minerals are present as weathering products in the fractures of the claystone, which are the major conduits for groundwater flow. Calcite and clay minerals are present in the rock matrix. Also present in the matrix is relict seawater that has not been flushed from the claystone. The seawater may provide its constituents, primarily Na^+ and Cl^-, to the groundwater by mixing or diffusion. The conceptual geochemical model for this system is based on the measured changes in groundwater concentration along the flow path and the known or expected reactive phases in contact with the groundwater. It consists of mineral equilibrium reactions with ferrihydrite, calcite, and gypsum; ion exchange with clays; and mixing or diffusion. The computer codes described in Chapter 5 can be used to calculate the amounts of reactants and products required to account for the change in groundwater composition.

Development of Multiple Hypotheses

In cases such as the marine claystone described above, sufficient knowledge of aquifer reactive phases may be available to develop an accurate conceptual model; however, in most cases data are not sufficient, and reactive phases must be proposed and tested with available data. It is important in the development of a conceptual model for a system to consider all the likely potential reactive phases and processes that might be influencing the system. This will generate several possible models that might explain some or all of the site data. For example, reducing conditions are often observed in aquifers not in contact with the atmosphere. The reducing conditions occur because dissolved oxygen has been consumed by an oxidation/reduction processes. Many reductants may exist in an aquifer. The reductant may be dissolved or suspended organic matter in the groundwater, organic matter in the aquifer solid phase, minerals containing reduced species of a wide variety of elements (e.g., Fe, Mn, S), or reducing gases (e.g., H_2S, CH_4, H_2). To focus on the actual process(es) causing reducing conditions, it is necessary to determine which of the reductants are present in a given system and to evaluate the products of the reaction of the various reductants with groundwater. For example, if organic matter is the reducing agent, then CO_2, and perhaps CH_4, will be produced by the reaction. An increase in CO_2 should affect the pH of the solution and the concentration of total inorganic carbon. These other effects can be investigated to decide whether organic carbon is a possible reactant in the system. Similarly, if pyrite was the reductant, it may drive many other reactions, as was seen for the case of the marine claystone.

Use of Saturation Indices in Developing Conceptual Models

The development of conceptual models is enhanced by saturation index calculations that can identify the equilibrium state of minerals (Section 3.2). Reactive minerals that have been identified in the aquifer, but which have a negative saturation index are expected to dissolve into the groundwater, increasing the solution concentrations of their components. In the marine claystone example, gypsum is undersaturated in the upgradient well; therefore it is expected to dissolve into the groundwater along the flow path, increasing calcium and sulfate concentrations. Reactive minerals that are in equilibrium with groundwater are probably limiting the solution concentration of one of their components. For the marine claystone, iron concentration is limited by the formation of ferrihydrite in the weathered zone where this mineral is in equilibrium with the water. A model that requires precipitation of a mineral with a saturation index less than zero to account for concentration decreases along the flow path is not realistic because concentrations are too low to allow precipitation. Halite, NaCl, is undersaturated in the marine claystone and would not be expected to limit the dissolved concentrations of either sodium or chloride.

4.3. GEOCHEMICAL MODELING METHODS

The two basic approaches to modeling aquifer geochemical interactions have been termed inverse and forward methods.[4] If the compositions of two groundwater samples along a flow path are known, the inverse method can be used to calculate possible water/rock/gas interactions that could produce the observed changes in composition. If the composition of groundwater is known at only one point and the investigator would like to predict changes in water composition due to natural geochemical processes along the flow path or by altering the groundwater system (e.g., *in situ* treatment of contaminants), then the forward modeling method is used.

The **inverse** method applies to situations where sufficient data are available to define the flow path and changes in groundwater composition along the path. It is assumed that reactive solids and gases along the flow path have produced the changes in composition, but the suite of reactive phases has not been identified nor is the amount of reactive phases that are dissolving or precipitating known. Figure 4-2 shows schematically the approach used by the inverse method. Models are developed consisting of suites of reactive phases that are known or conjectured to occur in the aquifer. Mass balance calculations are made to account for the change in solution composition attributable to the dissolution or precipitation of the reactive phases. The solution of the mass balance equations for a particular suite of reactive phases produces a reaction of the form:

Initial water composition + Reactants = Final water composition + Products

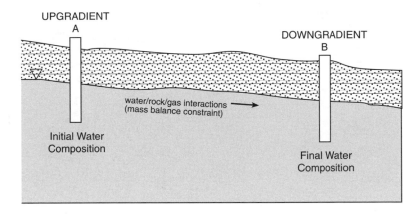

Initial composition at A + reactants =
Final composition at B + products

Figure 4-2   Inverse method of geochemical modeling.

This reaction accounts for the changes in water composition by the amount (in moles per kilogram of water) of reactants that must dissolve into the water and products that must precipitate from solution as the water flows from an upgradient well (initial composition) to a downgradient well (final composition).

Many potential conceptual models of different suites of reactive phases may account for the observed changes in solution concentration along a flow path, and it is the challenge of the modeler to eliminate unrealistic models. Because the inverse method is based on mass balance calculations only and does not inherently consider thermodynamic constraints on the system, the model may predict that a mineral is precipitating when concentrations in solution are not high enough to allow for the mineral to form. The calculation of saturation indices for the reactive phases is necessary to show that a particular model is realistic. Additional data from the system will also allow the modeler to reduce the number of potential models that explain the changes in water composition. For example, if information is known about the stable isotope distribution in the groundwater and the reactive phases, then isotope balance can be used with mass balance to refine the reaction model. The most commonly used mass balance computer codes that can be used for inverse modeling are BALANCE[5] and its successor NETPATH.[6] Computer codes for modeling calculations are described in Chapter 5.

The **forward** method of geochemical modeling applies to situations where data may only be available from one point along the flow path. In this case a prediction of the composition at a downgradient location is desired or the investigator wishes to know how the aquifer system will respond to the addition of a reactant or some other change in environmental condition (for example, temperature or P_{CO_2}). The forward methodology is shown in Figure 4-3. As with the inverse method, it is assumed that reactive solids and gases produce the changes in composition, but it is not known what the suite of reactive phases are nor the amounts of these phases that are dissolving or precipitating. In the forward method, reactants are added to the starting solution. This method uses equilibrium constants for all the potential reactions allowed by the model; therefore reactants will dissolve if they are undersaturated and products may be formed if adding the reactant causes a mineral to become saturated in the solution. The result of the model calculations once again consists of a suite of reactive phases; however, their reactivity is constrained

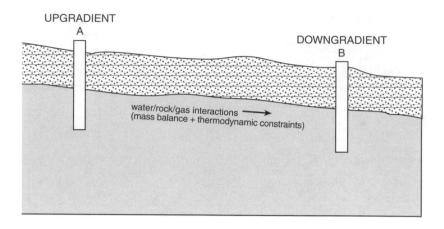

Initial composition at A + reactants = Predicted composition at B + products

Figure 4-3   Forward method of geochemical modeling.

by mass balance and thermodynamic equilibrium. In terms of an overall chemical reaction, the forward method can be viewed as follows:

Initial solution + Reactants = Predicted solution + Products

The amount of reactants (in moles per kilogram water) dissolving into the water and the amount of products precipitating from solution are calculated, as well as the resulting solution constrained either by mass of reactants added or the solubility of the mineral products. For an aquifer geochemical system, the forward model can be validated by comparing the composition of water in a downgradient well with the predicted composition. In the case of results of a laboratory treatability study, the results of testing can be compared with the model predictions. If the prediction does not match with actual field or lab data, then either the suite of reactive phases is not appropriate or slow reaction rates have not allowed the system to come to equilibrium.

In addition to simulating the natural system, the forward method of geochemical modeling is appropriate to use in predicting the movement of contaminants and in remediation design. In the case of simulating contaminant movement, the conceptual model starts with the natural system in equilibrium between groundwater and reactive minerals in the aquifer. The addition of a contaminant to this system creates a disequilibrium that produces reactions that progress in a direction to reduce the disequilibrium, ultimately establishing a new equilibrium condition. The forward model predicts this new equilibrium and distributes the contaminant between the mobile groundwater phase and the immobile solid phases. For restoration design, various chemical techniques of treating the aquifer system or, for pump and treat methods, the groundwater alone can be simulated by the forward method. For instance, neutralization of an acidic groundwater or waste stream can be simulated by adding increments of calcite or some other neutralizing agent to the aquifer or water. Secondary reactions such as gypsum precipitation may also be included in these models. Forward modeling of neutralization allows estimates to be made of the amount of calcite required to neutralize the solution to a given pH value and the amount of byproduct solids formed. Models of restoration should be validated by laboratory experiments and small scale field tests. The most commonly used computer codes for forward-reaction modeling are MINTEQ[7-9] and PHREEQE.[10] The capabilities of these modeling codes are described in Chapter 5.

REFERENCES

1. Thomas, J.M., Welch, A.H., and Preissler, A.M., Geochemical evolution of ground water in Smith Creek Valley — a hydrologically closed basin in central Nevada, U.S.A., *Appl. Geochem.,* 4, 493–510, 1989.
2. Plummer, L.N. and Back, W., The mass balance approach: application to interpreting the chemical evolution of hydrologic systems, *Am. J. Sci.,* 280, 130–142, 1980.
3. Henderson, T., Geochemical reduction of hexavalent chromium in the Trinity Sand Aquifer, *Ground Water,* 32, 477–486, 1994.
4. Plummer, L.N., *Geochemical Modeling: A Comparison of Forward and Inverse Methods,* National Water Well Association, Banff, Alberta, Canada, 1984.
5. Parkhurst, D.L., Plummer, L.N., and Thorstenson, D.C., *BALANCE — A Computer Program for Calculation of Chemical Mass Balance,* U. S. Geological Survey, 1982.
6. Plummer, L.N., Prestemon, E.C., and Parkhurst, D.L., *An Interactive Code (NETPATH) for Modeling Net Geochemical Reactions Along a Flow Path,* U.S. Geological Survey, 1991.
7. Felmy, A.R., Girvin, D.C., and Jenne, E.A., *MINTEQ: A Computer Program for Calculating Geochemical Equilibria,* U.S. Environmental Protection Agency, 1983.
8. Brown, D.S. and Allison, J.D., *MINTEQA1, Equilibrium Metal Speciation Model: A User's Manual,* U.S. Environmental Protection Agency, 1987.
9. Allison, J.D., Brown, D.S., and Novo-Gradac, K.J., *MINTEQA2/PROEDFA2: A Geochemical Assessment Model for Environmental Systems,* U.S. Environmental Protection Agency, 1991.
10. Parkhurst, D.L., Thorstenson, D.C., and Plummer, L.N., *PHREEQE — A Computer Program for Geochemical Calculations,* U.S. Geological Survey, 1980.

5
COMPUTER CODES
FOR GEOCHEMICAL MODELING

The process of geochemical modeling is shown in the flowsheet presented in Figure 5-1. The conceptual geochemical model discussed in Chapter 4 is based on a knowledge of the types of chemical reactions that can occur in the environment and specific data on the site being modeled. This conceptual geochemical model is combined with a computer code capable of performing all the important calculations for these reactions that impact mass transfer between the phases in the system. The combination of a site conceptual model with an appropriate computer code produces a geochemical model of the site that can be used to understand existing conditions and predict the response of the system to a change in conditions. This chapter describes some of the commonly used computer codes available for performing these calculations. Geochemical modeling codes can be divided into three basic types: mass balance, equilibrium speciation/saturation, and equilibrium mass transfer.

5.1. MASS BALANCE MODELING CODES

Codes that perform only mass balance calculations and do not incorporate equilibrium considerations are most appropriate for conceptual models where data are available for two points along the flowpath or at the beginning and end of an experiment. In the case of groundwater modeling, mass balance codes account for changes in concentrations of dissolved components of the groundwater between the two locations in terms of solid, gas, and/or solution (mixing) phase interactions with the groundwater. Components of the groundwater are the individually measured dissolved constituents such as Ca^{2+}, Mg^{2+}, and SO_4^{2-}. For each component of the groundwater, one or more reactive phases (solids, solutions) are defined that might account for changes in groundwater composition along the flow path. The number of reactive phases must not exceed the number of components or the Phase Rule (Section 1.4) will be violated. The code solves the set of equations relating changes in concentration of the components to possible dissolution, precipitation, or mixing of the phases containing the components. The equations for the components are of the following form[1]:

$$\Delta m_{T,k} = \sum_{p=1}^{P} \alpha_p b_{p,k} \qquad k = 1, j \qquad (5-1)$$

In this equation, $\Delta m_{T,k}$ is the change in total (T), dissolved concentration of the kth component along the flow path, which equals the sum of the amount (α_p, in moles/kg water) of all the phases (p) containing the kth component that dissolve into or precipitate from the groundwater. The term $b_{p,k}$ is the stoichiometric coefficient of the kth component in the pth mineral. For example, if carbonate is a component of interest that changes in concentration along the flow path and the conceptual model includes calcite ($CaCO_3$), dolomite [$CaMg(CO_3)_2$], and CO_2(gas) as potentially reactive phase, the mass balance equation for carbonate would be as follows:

$$\Delta m_{T,CO_3} = \alpha_{calcite} + 2\alpha_{dolomite} + \alpha_{CO_2} \qquad (5-2)$$

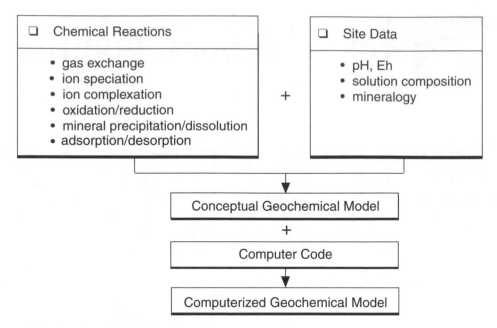

Figure 5-1   Geochemical modeling flow chart.

The change in the concentration of carbonate between two points along the flow path according to this model equals the amount of calcite and dolomite (in moles per kilogram of water) that dissolve into (α_p positive) or precipitate from (α_p negative) the groundwater plus the amount of CO_2 gas that dissolves into the water or outgases from it. If a phase is dissolving into the groundwater, then the mass of the component in solution will increase and α_p will be positive. If a phase is precipitating from the groundwater, then the mass of the component in solution will decrease and α_p will be negative. The amount of dolomite in the equation above is multiplied by a factor of two because each mole of dolomite contains two moles of carbonate. Additional mass balance relationships for calcium and magnesium for the selected reactive phases are as follows:

$$\Delta m_{T,Ca} = \alpha_{calcite} + \alpha_{dolomite} \tag{5-3}$$

$$\Delta m_{T,Mg} = \alpha_{dolomite} \tag{5-4}$$

The three mass balance relationships for the components can be solved for $\alpha_{calcite}$, $\alpha_{dolomite}$, and α_{CO_2} because there are three equations (5-2, 5-3, and 5-4) and three unknowns. The solution of the mass balance equations is the following reaction:

Initial water composition + Reactant phases = Final water composition + Product phases

Table 5-1 provides an example of a mass balance calculation using these components and reactive phases. For the conditions modeled, the interactions between phases along the flow path produce the following reaction:

$\{$Initial Water Composition$\}$
$\{$1 mmole/kg Ca^{2+} + 1 mmole/kg Mg^{2+} + 0.5 mmole/kg $CO_3^{2-}\}$ + 1 mmole/kg dolomite =

$\{$Final Water Composition$\}$
$\{$1.5 mmole/kg Ca^{2+} + 2 mmole/kg Mg^{2+} + 1.25 mmole/kg $CO_3^{2-}\}$ + 0.5 mmole/kg calcite
+ 0.75 mmole/kg $CO_2(g)$

TABLE 5-1.

MASS BALANCE CALCULATIONS FOR CARBONATE AQUIFER

Groundwater data	Upgradient well	Downgradient well	Δm_T
Ca (mmoles/kg)	1.0	1.5	0.5
Mg (mmoles/kg)	1.0	2.0	1.0
CO_3^{2-} (mmoles/kg)	0.5	1.25	0.75

Selected reactive phases	Components		
	Ca	Mg	CO_3^{2-}
Calcite	1		1
Dolomite	1	1	2
$CO_2(g)$			1

Mass balance equations

$\Delta m_{T,Ca} = \alpha_{calcite} + \alpha_{dolomite} = 0.5$

$\Delta m_{T,Mg} = \alpha_{dolomite} = 1.0$

$\Delta m_{T,CO_3} = \alpha_{calcite} + 2\alpha_{dolomite} + \alpha_{CO_2} = 0.75$

Mass balance solution

$\alpha_{calcite} = -0.5$ mmole/kg

$\alpha_{dolomite} = 1$ mmole/kg

$\alpha_{CO_2} = -0.75$ mmole/kg

Flow path reaction

Upgradient well water + 1 mmole/kg dolomite = downgradient well water + 0.5 mmole/kg calcite + 0.75 mmole/kg CO_2

To account for the changes in water composition given the selected reactive phases, 1 mmole of dolomite must dissolve into each kilogram of water while 0.5 mmoles of calcite must precipitate and 0.75 mmoles of CO_2 must outgas.

BALANCE[1] is the original mass balance computer code. In addition to calculating mass balance on components, it can model redox reactions by keeping track of electron transfer, perform isotope balance calculations, and allow for mixing of two water types to produce a final water composition. Because the reactions are not constrained by equilibrium considerations, BALANCE may predict that a mineral is precipitating from the water along the flowpath when the saturation index of the mineral is less than zero. This model would not be considered plausible. BALANCE modeling results must be checked for consistency with saturation index calculations.

The capabilities of BALANCE have been significantly enhanced by its successor NETPATH[2]. In addition to mass balance capabilities, NETPATH includes a preprocessor to develop input files and a modified version of the program WATEQF[3], which calculates the distribution of species (as a result of aqueous complexation) and saturation indices. The availability of saturation indices for the possible reactive phases enhances the selection of plausible reaction models. Rather than have the user develop separate suites of reactive phases and run them independently with the code as done with BALANCE, NETPATH allows the user to specify as many reactive phases as desired and the code develops and runs each reaction model automatically. Limitations can be placed on the types of models reported so that only the more plausible ones are retained. NETPATH can model mixing of two waters, as well as evaporation and dilution. In addition, if isotope data are available, it can predict carbon, sulfur, and strontium isotopic compositions, including radiocarbon dating. Several examples of the types of simulations that can be performed by NETPATH are included in the program documentation. The simulations include weathering of silicate rocks, evaporation of lake water, mixing and reaction of acid mine drainage, and application of carbon and sulfur isotope data to reaction modeling. The BALANCE and NETPATH codes are available from the following address:

> U.S. Geological Survey
> 437 National Center
> 12201 Sunrise Valley Dr.
> Reston, VA 22092
> Attn: O. A. Halloway
> 703-648-5695
> http://h20.usgs.gov/software/geochemical.html

5.2. EQUILIBRIUM SPECIATION/SATURATION MODELING CODES

To evaluate the equilibrium state of a solution with respect to mineral and gas phases, a method of calculating aqueous complexation is required. This is necessary because equilibrium is related to the activity (effective concentration) of the dissolved species, and the activity of a species in a given solution is a function of complexation reactions involving that species. If the activities of the dissolved species have been calculated, then equilibrium can be evaluated with respect to mineralogy, oxidation/reduction, gas exchange, and other interactions between the phases present in an aquifer.

Input to an aqueous speciation code consists of the field and laboratory data collected for a water sample. As discussed in Chapter 4 on the development of conceptual models, the type of data must be sufficient to allow for appropriate calculations to be made of all the important processes considered in the model. For example, if redox processes are important in the aquifer, either the redox potential must be measured or the concentrations of separate redox states of an element must be determined. If redox potential is both measured by the platinum electrode and calculated from the measured concentrations of two redox states of an element, then a comparison will show if the measured and calculated Eh are consistent. For modeling purposes, certain parameters must always be measured. These include temperature, pH, and concentrations of the major constituents in the water sample.

Aqueous speciation codes take the input data and distribute the total mass of each component in solution among its various aqueous species. This distribution is accomplished using the mass action relationships (formation reactions) for the various species in terms of the basic components of the species. For example, if Fe^{3+}, H^+, and H_2O are selected as components of the groundwater, then a reaction can be written to form the Fe(III) species $Fe(OH)^{2+}$ as follows:

$$Fe^{3+} + H_2O \leftrightarrow Fe(OH)^{2+} + H^+ \quad \left(K = 10^{-2.19}, 25°C \right) \tag{5-5}$$

This is a mass action expression of the reaction to form $Fe(OH)^{2+}$. The equilibrium constant for this reaction can be used to calculate the activity of $Fe(OH)^{2+}$ from its components:

$$\frac{\left(a_{Fe(OH)^{2+}} \right)\left(a_{H^+} \right)}{\left(a_{Fe^{3+}} \right)\left(a_{H_2O} \right)} = K = 10^{-2.19} \tag{5-6}$$

$$a_{Fe(OH)^{2+}} = 10^{-2.19} \frac{\left(a_{Fe^{3+}}\right)\left(a_{H_2O}\right)}{\left(a_{H^+}\right)} \tag{5-7}$$

The activity of water may be set equal to one because it is close to unity in dilute water or it may be calculated from the concentration of dissolved species. The activity of hydrogen is derived from the measurement of solution pH ($= -\log a_{H^+}$). The activity of Fe^{3+} is calculated from mass balance on total dissolved iron and the mass action relationships for all the iron species. For example, if the total dissolved iron concentration ($m_{Fe,T}$) was present in solution as only the Fe^{3+} and $Fe(OH)_2^+$ species, then there would be two equations, mass balance (5-8) and mass action (5-9), that can be solved to calculate the activity of Fe^{3+}:

$$m_{Fe,T} = m_{Fe^{3+}} + m_{Fe(OH)_2^+} \tag{5-8}$$

$$a_{Fe(OH)^{2+}} = 10^{-2.19} \frac{\left(a_{Fe^{3+}}\right)\left(a_{H_2O}\right)}{\left(a_{H^+}\right)} \tag{5-9}$$

The concentration (m) of a species is related to activity by its activity coefficient (γ), therefore

$$a_{Fe(OH)^{2+}} = \gamma_{Fe(OH)^{2+}}\left(m_{Fe(OH)^{2+}}\right)$$

$$a_{Fe^{3+}} = \gamma_{Fe^{3+}} m_{Fe^{3+}}$$

Substituting these relationships into Equation 5-9 and combining Equations 5-9 with 5-8 gives the following

$$m_{Fe,T} = m_{Fe^{3+}} + \left(10^{-2.19}\right)\frac{\left(\gamma_{Fe^{3+}} m_{Fe^{3+}}\right)\left(a_{H_2O}\right)}{\left(\gamma_{Fe(OH)^{2+}}\right)\left(a_{H^+}\right)} \tag{5-10}$$

Rearranging produces

$$m_{Fe^{3+}} = \frac{m_{Fe,T}}{1 + \left(10^{-2.19}\right)\dfrac{\left(\gamma_{Fe^{3+}}\right)\left(a_{H_2O}\right)}{\left(\gamma_{Fe(OH)^{2+}}\right)\left(a_{H^+}\right)}} \tag{5-11}$$

For this simplified example, the concentration of uncomplexed ferric iron in solution ($m_{Fe^{3+}}$) can be calculated from Equation 5-11 using the measured total iron concentration ($m_{Fe,T}$), the groundwater pH, and the ionic strength of the solution to calculate the activity coefficients $\gamma_{Fe^{3+}}$ and $\gamma_{Fe(OH)^{2+}}$. From its concentration ($m_{Fe^{3+}}$), the activity of Fe^{3+} can be calculated [$a_{Fe^{3+}} = (\gamma_{Fe^{3+}})(m_{Fe^{3+}})$] and used in Equation 5-9 to calculate the activity of the ferric iron species $Fe(OH)^{2+}$. Although there are actually many more species of ferric iron that must be considered in calculating the complete iron species distribution, the method of calculation for all the Fe(III) species is identical. For each additional species there is another mass action relationship and another species added to the mass balance on iron. Although the calculation method is seemingly straightforward, it can become very extensive if many species are involved, which is typical of a groundwater solution. For example, the actual mass balance equation for the iron/water system is as follows:

$$Fe_T = Fe(II) + Fe(III) = \left[Fe^{2+} + FeOH^+ + Fe(OH)_2^0 + Fe(OH)_3^- \right] +$$

$$\left[Fe^{3+} + FeOH^{2+} + Fe(OH)_2^+ + Fe(OH)_3^0 + Fe(OH)_4^- + Fe_2(OH)_3^{4+} + Fe_3(OH)_4^{5+} \right] \tag{5-12}$$

There is a mass action equation with thermodynamic data for each of these species of iron.

Also complicating the process is the fact that the calculations are interrelated. The calculated activities of the species affect the concentrations of the ions, which changes the ionic strength. The change in the ionic strength alters the activity coefficients and ultimately the activities, which must then be recalculated. However, this type of calculation is very amenable to computer processing, hence the development of computer codes for determining the distribution of aqueous species.

Once the activities of the dissolved species have been calculated for a water sample, other equilibrium relationships can be evaluated. Saturation indices [= log (ion activity product)/$K_{mineral}$] can be calculated for all minerals that can be formed from the components of the solution. For example, if the activities of ferrous and ferric iron are calculated for a solution, it is possible to determine the saturation indices of ferrihydrite (Fe(OH)$_3$, goethite (α-FeOOH), hematite (α-Fe$_2$O$_3$), maghemite (γ-Fe$_2$O$_3$), magnetite (Fe$_3$O$_4$), lepidocrocite (γ-FeOOH), and any other Fe(II) or Fe(III) oxyhydroxide mineral for which equilibrium constants are available. The partial pressure of a gas in equilibrium with the solution can be calculated for all gases that can be formed from the components of the solution. If carbonate is present in water, the theoretical partial pressure of CO$_2$ gas in equilibrium with the water can be calculated from the following reaction and equilibrium relationship:

$$CO_3^{2-} + 2H^+ \leftrightarrow CO_2(g) + H_2O \quad K = 10^{18.16} \tag{5-13}$$

$$P_{CO_2} = \left(10^{18.16} \right) \frac{\left(a_{CO_3^{2-}} \right)\left(a_{H^+} \right)^2}{\left(a_{H_2O} \right)} \tag{5-14}$$

If the concentrations of two redox states of an element have been measured in the water [e.g., Fe(II) and Fe(III)], the speciation calculations can determine the activities of the two species from which a redox potential can be calculated using the Nernst Equation. As discussed in Chapter 2, the Nernst Equation for the Fe(II)/Fe(III) redox couple is as follows:

$$Eh\,(mV) = 0.77\,V + 0.059\log \frac{a_{Fe^{3+}}}{a_{Fe^{2+}}} \tag{5-15}$$

Therefore, the theoretical Eh value can be calculated from the derived values for the activities of Fe^{3+} and Fe^{2+}.

The original, publicly available equilibrium aqueous speciation/saturation computer program was the WATEQ code.[4] WATEQ can perform speciation/saturation calculations for all of the major components of groundwater and many of the trace components. Its thermodynamic database contains equilibrium constants and enthalpy values (for temperature corrections of K) for several hundred species and minerals. Improvements to WATEQ have primarily added thermodynamic data for additional elements and increased the speed of execution. The FORTRAN version of the code (WATEQF) is available as part of NETPATH, which runs on personal computers. Numerous other speciation/saturation codes have been developed over the years. These codes were compared by Nordstrom et al.[5] who found some significant differences in the results of the calculations primarily because of differences in the thermodynamic databases. Other factors that lead to discrepancies are methods of calculating activity coefficients, redox assumptions, temperature corrections, alkalinity corrections, and number of aqueous complexes considered.

As shown in Table 5-2, the WATEQ family of codes and those similar to it can perform the following calculations: ion speciation, adjust equilibrium constant for the temperature of the system, redox equilibrium, gas phase equilibrium, and saturation indices. The saturation indices are reported for all possible minerals in the database that could form given the input components of the groundwater. It is up to the modeler's judgement to choose appropriate reactive minerals from the list. These codes are very useful in confirming the plausibility of proposed reaction models. For instance, the calculation may show that the precipitation reaction proposed by a mass balance model is not realistic because the mineral is undersaturated in the system.

TABLE 5-2.

MODELING CAPABILITIES OF EQUILIBRIUM CODES

Capabilities	WATEQ	MINTEQA2	PHREEQC
Ion speciation	★	★	★
K at system temp.	★	★	★
Redox equilibrium	★	★	★
Gas phase equilibrium	★	★	★
Saturation indices	★	★	★
Database	★	★ ★	★
Mass transfer		★	★
Ion exchange		★	★
Adsorption isotherm		★	
Surface complexation		★	☆
Titration			★
Mixing			★
Reaction path			★
Advective transport			☆

Although the WATEQ-type codes are very good at calculating saturation indices, they do not have the ability to transfer mass between phases, such as the dissolution of a mineral into solution to reach equilibrium. This ability is found in the equilibrium mass transfer codes.

5.3. EQUILIBRIUM MASS TRANSFER MODELING CODES

To model changing water composition along a flow path or predict the impact of changing a system variable (such as temperature or pH) on solution composition, the calculation method must be able to transfer mass between the various phases in the system. As described in Section 5.1, the mass balance codes transfer mass, but do not incorporate an equilibrium constraint. This section describes the mass transfer methodology that does include chemical equilibrium. Figure 5-2 shows a flow path for a typical equilibrium mass transfer calculation. The composition of the groundwater is entered into the aqueous speciation subroutine, which calculates the concentrations and activities of all possible species that can be formed from the components of the solution. This subroutine uses the mass balance and mass action equations to solve for activities in the same manner as described in Section 5.2 for the equilibrium speciation codes. The activities are then used to calculate ion activity products for all possible minerals and compare the IAPs to the equilibrium constants for the minerals.

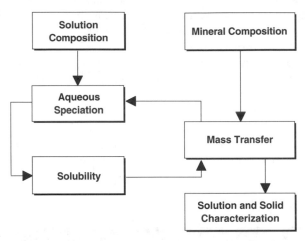

Figure 5-2　Equilibrium mass transfer calculations.

At this point the modeler can stipulate the type of mass transfer reactions allowed or let the code automatically equilibrate with the most stable phases. Because the databases of the codes typically include minerals that will not equilibrate in an aquifer environment, it is best that the reactive minerals be selected by the modeler. Mass transfer between phases may consist of mineral dissolution/precipitation, adsorption/desorption and/or gas dissolution/exsolution. As mass is transferred, the mass balance on components in solution changes and aqueous speciation must be calculated again. When enough mass has been transferred to reach equilibrium, or some other preset condition, calculations are stopped and the output shows the amount of mass transfer that occurred and the resulting solution composition.

The calculations performed by the equilibrium mass transfer codes are identical to those of the speciation/saturation codes with the addition of the ability to transfer mass from one phase to another. The constraints on this mass transfer that can be imposed in the model of a system allow for a wide variety of interactions to be simulated. The solution may be allowed to equilibrate with one or more minerals either by dissolving or precipitating them. For the case where a mineral is present in limited supply in an aquifer (e.g., a calcite cement or ferrihydrite mineral coating), the modeler can stipulate the amount of the mineral (in moles per kilogram of water) present in the system. The amount of adsorption and desorption onto a mineral can be limited by the concentration of the mineral in the solid phase of the aquifer. The partial pressure of a gas can be fixed in a system allowing gas to dissolve into or exsolve from the solution in response to other reactions between the solution and solid phases.

Reaction path simulations are a subset of equilibrium mass transfer calculations. In a reaction path simulation, incremental additions of one mineral are made to a solution, and minerals that become saturated are allowed to precipitate. This type of simulation differs from the single-step equilibrium calculations by providing information about the evolutionary path of the water as each increment of reactant is added to solution. For example, feldspar minerals are common in many different types of aquifers. They dissolve slowly relative to groundwater residence times. As a consequence, they provide their components to the water, but rarely equilibrate with the solution. Secondary minerals (silicon and aluminum oxides and hydroxides and clays) containing the components of the feldspars may equilibrate with the water. Because the feldspars do not themselves equilibrate with the water, it would not be appropriate to specify a feldspar in equilibrium with the water, but it may be important to model the dissolution of the feldspar and allow the secondary minerals to precipitate if they become saturated. For this type of simulation, a code that can simulate reaction path is preferable to a single-step equilibrium code.

Two of the more commonly used and readily available equilibrium mass transfer codes are MINTEQ and PHREEQE. MINTEQ[6] was originally developed by Battelle Pacific Northwest Laboratory for the U.S. Environmental Protection Agency by combining the WATEQ code with MINEQL,[7] which has an efficient mass transfer routine. MINTEQ has been enhanced several times by the EPA by adding thermodynamic data, a preprocessor for developing input files, and options for capturing selected output.[8,9] PHREEQE (PH REdox EQuilibrium Equations) was developed by the U.S. Geological Survey [10] and has been enhanced in the PHREEQC[11] version by the ability to model surface complexation with the diffuse double-layer model and advective transport. Table 5-2 shows the general capabilities of the MINTEQ and PHREEQE families of computer codes. Both codes can model simple ion exchange and surface complexation using the diffuse double-layer approach. At this time, only MINTEQ has algorithms to model adsorption/desorption using the isotherm approach (linear Kd, Freundlich, and Langmuir), as well the constant capacitance and triple-layer surface complexation models. PHREEQE has several capabilities not found in MINTEQ. PHREEQE can simulate titrations, mixing of two solutions, reaction path and advective transport. Because MINTEQ and PHREEQE use the ion association theory in calculating aqueous speciation, they are limited to modeling solutions with an ionic strength of less than about 0.5. This is not a severe limitation for most groundwaters, which rarely exceed an ionic strength of 0.1; however, these codes are not appropriate for modeling evaporation processes, brines, or other solutions with high ionic strength. PHREEQE has been modified to simulate high ionic strength solutions by incorporating the Pitzer virial coefficient equations for calculating activity.[12] This version of PHREEQE is called PHRQPITZ.[13]

The latest version of the MINTEQ computer code is available from:

Center for Exposure Assessment Modeling (CEAM)
U.S. Environmental Protection Agency
Office of Research and Development
College Station Road
Athens, Georgia 30613-0801
706-546-3154

PHREEQE and PHRQPITZ are available from:

U.S. Geological Survey
437 National Center
12201 Sunrise Valley Dr.
Reston, VA 22092
Attn: O. A. Halloway
703-648-5695
http://h20.usgs.gov/software/geochemical.html

Training classes for these codes are given periodically by the U.S. EPA, the U.S. Geological Survey, and professional organizations, including the National Ground Water Association and Environmental Education Enterprises.

5.4. MASS TRANSPORT

The geochemical modeling codes based on mass balance and mass action equations adequately simulate the chemical interactions in a groundwater system; however, most do not simulate the physical transport processes of advection, dispersion, and diffusion. The geochemical codes model the exchange (transfer) of mass between phases, but not the movement (transport) of dissolved mass in response to hydraulic or concentration gradients. Because of the importance of chemical interactions on mobility, the transport codes that are designed to model the physical processes of flow and transport usually include a retardation term for reactive (nonconservative) constituents that do not move at the flow rate of the groundwater. Simple retardation factors may adequately simulate the chemical processes affecting movement if the chemistry of the aquifer (both solution and solid phases) is uniform along the flow path, if mineral precipitation is not a factor limiting solution concentration, if adsorption/desorption equilibrium is achieved and fits a linear isotherm, and if the solute is not impacted by biodegradation. In many situations of interest these assumptions are not met, and mass transport modeling with simple retardation does not provide a realistic simulation of the movement of a reactive constituent.

The coupling of a complete geochemical modeling code such as MINTEQ or PHREEQE with a finite difference or finite element transport code has not been considered a feasible approach to mass transport modeling with extensive geochemical reactions because of the computationally intensive requirement of modeling the geochemistry at each node of the grid. An approach that has lent itself to solving complex chemical interactions is the "mixing cells in series" concept.[14] In this approach, a packet of water influenced by physical and chemical processes is followed along a flow path. Physical processes affect the rate of flow and transfer of mass by dispersion and diffusion with adjacent packets. Geochemical processes allow for mass transfer between the solution and solid phases and a gas phase. The physical and geochemical transport processes are calculated separately and then combined to couple the processes.

The most readily available hydrogeochemical transport codes are PHREEQC and PHRE-EQM. PHREEQC was described in Section 5.3. The PHREEQM code is a combination of PHREEQE with a one-dimensional transport code with dispersion and diffusion. It includes all the geochemical modeling capabilities of PHREEQE with a mixing cell flowtube approach to physical transport. Guidelines for using PHREEQM and several example applications are provided in Appelo and Postma.[14] The code can be obtained from the following address:

A.A. Balkema Publishers
P.O. Box 1675
3000 BR Rotterdam
Netherlands
[telefax: (+31 10) 4135947]

Other mass transport codes that provide sophisticated geochemical modeling include TRANQL,[15] which can simulate surface complexation with the constant capacitance model, HYDROGEOCHEM,[16] which contains the triple layer model, and MINTRAN,[17] which has the full range of MINTEQA2 geochemical modeling capabilities coupled with the transport code PLUME2D.

5.5. GEOCHEMICAL MODELING LIMITATIONS

The ability of a geochemical model to accurately simulate natural systems is limited by the factors listed in Table 5-3. The model of a system is based on the information available about that system, therefore the field and laboratory-derived data for each phase of the system must be accurate and complete. A discussion of data needs and a list of the minimum amount of data are presented in Section 4.1. The equilibrium modeling codes also require accurate and complete thermodynamic data for all of the important aqueous species, and they require an accurate method of modifying the standard data to site conditions of temperature and ionic strength. Most of the commonly used codes are limited to temperatures less than 100°C and an ionic strength <0.5. The EQ3 code[18] can be used to model high temperature and pressures and PHRQPITZ[13] can be used to model high ionic strength solutions.

TABLE 5-3.

GEOCHEMICAL MODELING LIMITATIONS

Limiting Factor	Scope
Field and laboratory data	Field parameters (temp., pH, Eh, etc.)
	Lab analysis (accuracy, completeness)
	Solid phase characterization
Thermodynamic data	Database (accuracy, completeness)
(aq. species, solids, gases)	Calculation of activity coefficients
	Calculation of K_T
Equilibrium constraint	Minerals, gases and redox state
Ionic strength of solution	I < 0.5: ion association theory
	I > 0.5 ion interaction theory
Range of geochemical processes	Ion complexation, mineral dissolution/precipitation, redox changes, gas phase equilibria, adsorption/desorption, solid solution

The commonly available modeling codes do not have the capability to simulate chemical reaction rates; therefore reactions that can equilibrate in the time frame under consideration, such as groundwater residence time, must be selected. This includes reactive minerals that can dissolve or precipitate in the allowed time, as well as redox reactions that can achieve equilibrium concentrations between the redox pairs of an element. Finally, the code that is selected to make the necessary calculations for the model must have the capability to simulate all the important processes in the system. For example, if trace metal adsorption is an important process, then MINTEQ would be the preferred code; whereas, if the intent is to simulate reaction path or mixing of waters, then PHREEQE would be a better choice. PHREEQC and PHREEQM are available to model coupled physical and chemical mass transport/transfer processes.

The model developed for a natural system should be validated by comparing the modeling results with field or laboratory data. The computer code can almost always provide results, but whether they correspond to reality can only be determined by comparing the output to actual data. Differences may be due to inaccuracies in the computational method or in the conceptual model. If the computational method has been verified by hand calculations or comparison with the results of another suitable computer code, then it is likely that the conceptual model is not adequate. The process of validating a geochemical model can lead to a greater understanding of the geochemical system and refinement of the conceptual model.

REFERENCES

1. Parkhurst, D.L., Plummer, L.N., and Thorstenson, D.C., *BALANCE — A Computer Program for Calculation of Chemical Mass Balance*, U.S. Geological Survey, 1982.
2. Plummer, L.N., Prestemon, E.C., and Parkhurst, D.L., *An Interactive Code (NETPATH) for Modeling Net Geochemical Reactions Along a Flow Path*, U.S. Geological Survey, 1991.
3. Plummer, L.N., Jones, B.F., and Truesdell, A.H., *WATEQF — A FORTRAN IV Version of WATEQ, a Computer Program for Calculating Chemical Equilibria of Natural Waters*, U.S. Geological Survey, 1976.
4. Truesdell, A. and Jones, B.F., *WATEQ — A Computer Program for Calculating Chemical Equilibria of Natural Waters*, U.S. Geological Survey, 1973.
5. Nordstrom, D.K. et al., in *Chemical Modeling in Aqueous Systems*, Jenne, E.A., Eds., American Chemical Society, Washington, DC, 1979, pp. 857–892.
6. Felmy, A.R., Girvin, D.C., and Jenne, E.A., *MINTEQ: A Computer Program for Calculating Geochemical Equilibria*, U.S. Environmental Protection Agency, 1983.
7. Westall, J., Zachary, J.L., and Morel, F.M.M., *MINEQL, a Computer Program for the Calculation of Chemical Equilibrium Composition of Aqueous Systems*, M. I. T., 1976.
8. Brown, D.S. and Allison, J.D., *MINTEQA1, Equilibrium Metal Speciation Model: A User's Manual*, U.S. Environmental Protection Agency, 1987.
9. Allison, J.D., Brown, D.S., and Novo-Gradac, K.J., *MINTEQA2/PROEDFA2, a Geochemical Assessment Model for Environmental Systems*, U.S. Environmental Protection Agency, 1991.
10. Parkhurst, D.L., Thorstenson, D.C., and Plummer, L.N., *PHREEQE — A Computer Program for Geochemical Calculations*, U.S. Geological Survey, 1980.
11. Parkhurst, D.L., *User's Guide to PHREEQC, a Computer Model for Speciation, Reaction Path, Advective Transport and Inverse Geochemical Calculations*, U.S. Geological Survey, 1995.
12. Pitzer, K.S., Thermodynamics of electrolytes. I. Theoretical basis and general equations, *J. Phys. Chem.*, 77, 268–277, 1973.
13. Plummer, L.N., Parkhurst, D.L., Fleming, G.W., and Dunkle, S.A., *A Computer Program Incorporating Pitzer's Equations for Calculation of Geochemical Reactions in Brines*, U.S. Geological Survey, 1988.
14. Appelo, C.A.J. and Postma, D., *Geochemistry, Groundwater and Pollution*, A.A. Balkema, Rotterdam, 1994.
15. Cederberg, G.A., Street, R.L., and Leckie, J.O., A groundwater mass transport and equilibrium chemistry model for multicomponent systems, *Water Resour. Res.*, 21, 1095–1104, 1985.
16. Yeh, G.-T. and Tripathi, V.S., *HYDROGEOCHEM: A Coupled Model of HYDROlogic Transport and GEOCHEMical Equilibria in Reactive Multicomponent Systems*, Oak Ridge National Laboratory, Oak Ridge, TN, 1990.
17. Walter, A.L., Frind, E.O., Blowes, D.W., Ptacek, C.J., and Molson, J.W., Modeling of multicomponent reactive transport in groundwater I. Model development and evaluation, *Water Resour. Res.*, 30, 3137–3148, 1994.
18. Wolery, T.J., *EQ3NR: A Computer Program for Geochemical Aqueous Speciation-Solubility Calculations*, Lawrence Livermore Laboratory, California, 1983.

PART II.
GENERAL GEOCHEMICAL APPLICATIONS

A knowledge of the basic geochemical processes and calculation methods described in Part I of this book can be used to develop appropriate sampling programs, understand the chemical processes affecting the mobility of compounds, and contribute to the design of effective remediation programs for contaminants in the subsurface. Chapter 6 provides an example of the general application of geochemistry to develop a geochemical model of a site that can be used to provide quantitative estimates of the impact of water/rock interactions on groundwater and solid phase composition of an aquifer. Subsequent chapters discuss the general impact of geochemistry on the movement of contaminants in the subsurface (Chapter 7) and the immobilization or removal of those contaminants by natural restoration, applied *in situ* methods, or chemical enhancements to the pump-and-treat remediation technique (Chapter 8). Chapter 9 describes the importance of geochemistry to the sampling process and discusses methods of ensuring that samples are representative of aquifer conditions.

GEOCHEMICAL MODELS

6

The geochemical model developed for a particular site is a simplification of the real world. It is not possible, or necessary, to include each and every constituent or process active in a system in your model. Some constituents may be relatively unreactive, while the reaction rates of some processes may be so slow that they do not significantly affect mass transfer in the system. For example, exchange of nitrogen gas between the atmosphere and water is usually neglected in a model because the dissolved nitrogen gas is not particularly reactive. In contrast, dissolved oxygen and carbon dioxide gases are reactive and may need to be included in a model of water in contact with the atmosphere or soil gas. The neutralization of acidic water by the primary silicate minerals in an aquifer is a slow process, which might be ignored in the geochemical model of an aquifer system with a groundwater residence time of less than a few hundred years and if other more reactive neutralizing minerals are present. Alternatively, neutralization by carbonate minerals should be part of the model if these minerals are present because they actively will actively participate in water/rock interactions.

This chapter describes several fairly simple models of both natural and human-impacted systems. The data required to model these systems are discussed stressing the implications of ignoring important data. The models are used to predict the impact of perturbations to the systems on mineral solubility and constituent mobility. Guidelines for developing conceptual geochemical models are described in Chapter 4, and the computer calculation methods for providing quantitative results of the model simulations are discussed in Chapter 5.

6.1. MOBILITY OF IRON IN THE NATURAL ENVIRONMENT

The mobility of iron can be used to illustrate some of the capabilities of geochemical modeling. Iron is a common constituent of many primary minerals (biotite, pyroxenes, and amphiboles), and it is a component of a number of reactive, secondary minerals (e.g., ferrihydrite, siderite, jarosite, and pyrite) that form under a wide range of pH and Eh conditions. The solubilities of these reactive minerals under aquifer conditions can limit the dissolved iron levels in groundwater and, consequently, the mobility of iron in the subsurface.

A hypothetical sequence of aquifer environments can be considered to show the application of geochemical modeling to understand the mobility of iron. The sequence of environments shown in Figure 6-1 includes the following zones: (1) high Eh, near-neutral pH; (2) low pH, high sulfate; 3) carbonate neutralization; (4) slightly acidic, low Eh without sulfide present and (5) low Eh with sulfide present. To simulate these environments it is necessary to know their pH and Eh values and the concentrations of the major ions in solution. The initial solution is a Ca–HCO_3 water type with measurable amounts of Mg, Na, K, Cl, and SO_4 (Table 6-1). (Other constituents, such as Si are present in the water, but they do not have an appreciable effect on iron mobility so they do not need to be included in the geochemical model.) The pH of the initial groundwater (Zone 1) is 7.5 and the Eh is +400mV. The water is in equilibrium with the iron mineral ferrihydrite [$Fe(OH)_3$], producing a total dissolved iron concentration of 8.5×10^{-9} moles/kg (0.0005 mg/L), which is present

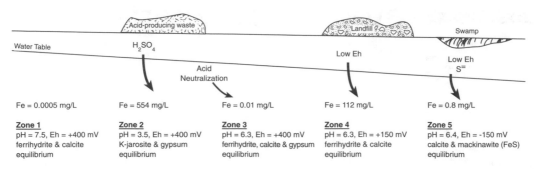

Figure 6-1   Environmental effect on iron mobility.

TABLE 6-1.

MODELING THE MOBILITY OF IRON

Solution parameters[a]	Zone I Near neutral pH Oxidizing[b]	Zone II Acidic, Oxidizing[c]	Zone III Neutralization w/calcite[d]	Zone IV Reducing w/o sulfide[e]	Zone V Reducing w/sulfide[f]
Fe(II)	8.00E-12	9.90E-03	7.80E-08	2.00E-03	1.40E-05
Fe(III)	8.50E-09	1.30E-07	1.20E-07	1.20E-07	2.20E-16
Fe(total)	8.51E-09	9.90E-03	1.98E-07	2.00E-03	1.40E-05
Ca	2.77E-03	3.80E-02	5.40E-02	5.20E-02	2.30E-02
HCO_3 +CO_3	2.77E-03	4.00E-02	9.10E-03	9.20E-03	1.00E-02
SO_4	1.50E-03	5.80E-03	4.60E-03	4.80E-03	1.70E-03
HS					2.10E-03
pH	7.5	3.5	6.3	6.3	6.4
Eh (mV)	400	400	400	150	−150

[a] Concentration units: moles/kg (constant concentration: Mg: 5E-4; Na: 1E-3; K: 2.5E-4; Cl: 1E-3).
[b] In equilibrium with calcite and ferrihydrite.
[c] In equilibrium with K-jarosite and gypsum.
[d] In equilibrium with calcite, gypsum, ferrihydrite. PCO_2 = 0.1 atm.
[e] In equilibrium with calcite and ferrihydrite. PCO_2 = 0.1 atm.
[f] In equilibrium with calcite and FeS. PCO_2 = 0.1 atm.

primarily as Fe(III) species in solution. The water also is in equilibrium with calcite. Under oxidizing, near-neutral pH conditions, the low solubility of ferrihydrite limits the dissolved iron concentration to very low levels.

In the acidic environment (Zone 2) where the pH is 3.5, ferrihydrite is much more soluble and will not limit the dissolved iron to such low levels. In the case where the acid environment is caused by sulfuric acid, it is likely that an iron sulfate mineral such as potassium jarosite [$KFe_3(SO_4)_2(OH)_6$] will limit the dissolved iron level. The simulation for the Zone 2 environment (Table 6-1) allows for the precipitation of K-jarosite and gypsum to provide limits on solution concentrations. The solubility of K-jarosite under the conditions of the acidic environment limits the iron concentration to 9.9×10^{-3} moles/kg (554 mg/L). Under these conditions iron is much more mobile than under neutral, oxidizing conditions.

If the acidic conditions are localized, the natural environment may neutralize the acid and produce pH values closer to neutral. This can be simulated with a geochemical model by equilibrating the solution from Zone 2 with calcite, which is a common neutralizing agent in natural systems. The results of the model calculations shown for Zone 3 are similar to Zone 1 except that the pH is 1.2 units lower in Zone 3. (A higher pH could have been simulated by allowing more calcite to neutralize the acid water and by allowing CO_2 to outgas.) Once again, ferrihydrite has been allowed to limit the iron concentration. At a pH of 6.3, ferrihydrite is much less soluble than at a pH of 3.5, but more soluble than at a pH of 7.5. The resulting total dissolved iron concentration is 1.98×10^{-7} moles/kg (0.01 mg/L).

Another environmental condition that can have a major effect on iron mobility is the redox potential (Eh) of the system. Iron occurs in the ferrous [Fe(II)] and ferric [Fe(III)] redox states, and the solubilities of minerals containing these forms of iron are sensitive to the Eh of the aquifer. Reducing conditions with respect to iron are shown in the simulation for Zone 4 in which the solution is in equilibrium with ferrihydrite at an Eh of 150 mV. Compared with more oxidizing conditions, ferrihydrite is relatively soluble at this Eh and near-neutral pH conditions. The resultant dissolved iron concentration is 2×10^{-3} moles/kg (112 mg/L) with most of the dissolved iron present as Fe(II). Reducing conditions may exist naturally in wetland areas or may be created by disposing of degradable organic matter in the soil such as at landfills.

As shown for Zone 4, if ferrihydrite limits iron concentrations in solution under reducing conditions, the dissolved iron can be fairly high. However, iron also forms Fe(II) minerals that may more severely limit iron concentrations in groundwater. Fe(II) sulfide minerals, such as mackinawite (FeS) and pyrite (FeS_2) are particularly insoluble under reducing conditions. In the simulation for Zone 5, redox equilibrium has been allowed for two sulfur redox states, S(-II) and S(VI). Also, the mineral mackinawite has been allowed to precipitate if it is oversaturated. As can be seen in Table 6-1, at an Eh of −150 mV the amount of iron in solution is lower than for the reducing conditions of Zone 4. This is because mackinawite has precipitated in Zone 5 and limited the dissolved iron level to 1.4×10^{-5} moles/kg (0.8 mg/L). The iron level would decrease further if a lower Eh was chosen because more of the sulfur would be present as S(-II), limiting the dissolved iron concentration in equilibrium with mackinawite to a lower level. At an Eh greater than −150 mV under near-neutral pH conditions, very little of the dissolved sulfur is present as S(-II); therefore mackinawite would be more soluble at higher Eh values, and the equilibrium iron concentration would be greater than 0.8 mg/L.

This modeling exercise with iron has shown the wide range of concentrations, hence mobilities, of iron within the pH/Eh ranges possible in the environment. It has also shown how geochemical modeling can be used to simulate these environments and estimate potential iron concentrations in solution under various scenarios of solution compositions and mineral equilibria. An example of the effect of redox potential and the presence of sulfide in an aquifer on dissolved iron concentrations is presented by Chapelle and Lovley[1] for the Middendorf Aquifer of South Carolina. Bennett et al.[2] document increased mobility of iron and other pH- and redox-sensitive elements due to contamination of a shallow aquifer by crude oil.

6.2. MODELING CALCITE MINERAL PRECIPITATION IN RESPONSE TO GROUNDWATER HEATING

Aquifer Thermal Energy Storage (ATES) is a concept developed by the U.S. Department of Energy to maximize the efficiency of power plants by running them all year at an optimal output level and storing excess energy as heat in groundwater. During periods of the year when more energy is needed than the plant can produce, the shortfall is made up by extracting energy from the groundwater. The concept is attractive because of the relatively high heat capacity of water and the potential simplicity of the system. The design includes supply and injection/recovery wells with a heat exchanger connecting the groundwater component of the system to the power plant (Figure 6-2). An important geochemical consideration in the siting of an ATES facility is the potential for mineral precipitation and plugging in response to heating of the groundwater. Carbonate minerals are less soluble at higher temperatures and could precipitate from groundwater initially at or near saturation with these minerals. Calcite is a common carbonate mineral that is at equilibrium with many groundwater types. Geochemical modeling can be used to evaluate the potential for

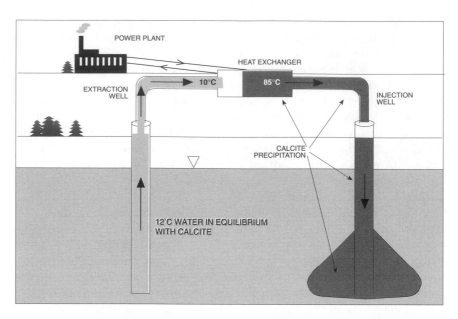

Figure 6-2   Aquifer thermal energy storage (groundwater heating cycle).

this mineral to precipitate and can provide an estimate of the mass and volume of mineral that might precipitate and would need to be taken into account in the design of an ATES system.

To model a groundwater system for mineral equilibria, a complete analysis of the major cations and anions is necessary because of their influence on ionic strength and complexation. In addition, the groundwater pH is required because of its importance to aqueous complexation, especially in the case of the carbonate system where the dominant species are strongly pH dependent (Figures 2-5 and 2-6). The redox potential is not a necessary parameter for modeling calcite equilibria, unless the system is very reducing and the dissolved concentration of iron, manganese or another redox-sensitive metal approaches the level of the major ions. The groundwater temperature is an important parameter because of the sensitivity of calcite solubility to temperature. Data compiled for an ATES test facility in St. Paul Minnesota[3] can be used to simulate the effect of heating on mineral equilibria. Table 6-2 lists the initial groundwater composition.

The groundwater at 12°C is close to equilibrium with calcite as shown by the saturation index for this mineral of –0.06. Because calcite is less soluble as temperature increases, it is likely that calcite will precipitate from solution during the ATES groundwater heating cycle. The calculated equilibrium carbon dioxide gas partial pressure is $10^{-2.13}$ atm. This is over 20 times the atmospheric P_{CO2} gas pressure ($10^{-3.5}$ atm); consequently, simulating the ATES system should include the possibility of venting carbon dioxide gas to the atmosphere and the effect of this loss of CO_2 on the solubility of calcite. The results of a MINTEQ simulation of heating this groundwater to 50°C, 75°C, and 99.9°C under closed and open conditions with respect to CO_2 gas are shown in Table 6-2. The saturation index for calcite was set equal to zero so that any oversaturated calcite would precipitate from solution. In the case of a system closed to the exchange of CO_2 gas with the atmosphere, increasing the temperature from 12°C to 50°C causes 3.1×10^{-4} moles of calcite to precipitate per kilogram of groundwater. This is equal to 26% of the initially available calcium in solution and 6% of the total carbonate. The system is designed to heat groundwater to over 100°C. At 99.9°C, 5.97×10^{-4} mole/kg calcite will precipitate, which is 60 mg/kg or 0.022 cc/kg. At a test flow rate of 360 gpm, calcite would precipitate at a rate of 30 cc/min or about 43 liters of calcite per day.

TABLE 6-2.

SIMULATION OF CALCITE PRECIPITATION DUE TO GROUNDWATER HEATING

Parameters	Initial[a]	CO₂ closed system			CO₂ open system			
		50°C	75°C	99.9°C	12°C	50°C	75°C	99.9°C
Ca (molal)	1.19E-03	8.80E-04	7.04E-04	5.93E-04	1.73E-04	4.70E-05	2.96E-05	2.76E-05
CO₃ (molal)	4.87E-03	4.56E-03	4.38E-03	4.27E-03	2.50E-03	2.10E-03	1.91E-03	1.63E-03
pH	7.46	7.16	6.94	6.71	8.6	8.74	8.71	8.6
SI Calcite	−0.06	0	0	0	0	0	0	
Amount pptd (mol/kg)		3.10E-04	4.86E-04	5.97E-04	1.02E-03	1.14E-03	1.16E-03	1.16E-03
Amount pptd (cc/kg)		0.011	0.018	0.022	0.038	0.042	0.043	0.043
Log PCO₂ (atm)	−2.128	−1.65	−1.45	−1.32	−3.5	−3.5	−3.5	−3.5
Amount degased (mol/kg)					1.34E-03	1.62E-03	1.80E-03	2.10E-03

[a] Temp. = 12°C; Mg: 8.7E-4 molal; Na: 2.4E-4; K: 6.9E-4; Cl: 2.6E-5; SO_4: 1E-4; SiO_2: 1.22E-4.

Data from Holm, T.R. et al., *Water Resour. Res.*, 23, 1005–1019, 1987.

Greater amounts of calcite precipitate from a system open to CO_2 gas exchange with the atmosphere. As shown in Table 6-2, temperature apparently has little impact on this precipitation because similar amounts (1.1×10^{-3} mol/kg) precipitate at 12°C and the elevated temperatures when the groundwater is allowed to equilibrate with carbon dioxide gas in the atmosphere. The loss of carbon dioxide gas from the groundwater allows the pH of the water to increase from 7.46 to 8.6, which lowers the solubility of the calcite a greater amount than the temperature increase. This type of modeling shows the importance of maintaining a closed system with respect to CO_2 gas to minimize calcite precipitation.

Geochemical modeling has shown the potential detrimental side effect of mineral precipitation when groundwater is heated in the ATES process or is allowed to degas CO_2. Modeling can also be used to evaluate alternative methods of eliminating calcite precipitation in the system. The most common methods of increasing mineral solubility are to (1) add a complexing agent to the solution that will lower the activity of one or more of the components of the mineral and (2) adjust the pH so that the activity of a pH-sensitive component of the mineral decreases. In the case of calcite, a complexing agent such as phosphate may be an answer to the problem of mineral precipitation because of phosphates' moderately strong complexing ability with calcium. The addition of an acid such as HCl might also be evaluated for its effect on carbonate speciation and calcite solubility.

Simulating the effect on calcite solubility of adding phosphate to the heated groundwater under closed system conditions is shown in Table 6-3. It can be seen that adding phosphate ties up some of the calcium in solution as calcium phosphate complexes (e.g., $CaPO_4^0$, $CaHPO_4^+$). This complexation decreases the amount of free calcium available in solution to form calcite, thereby increasing the solubility of calcite and decreasing the amount of calcite that precipitates. The simulation shows that the amount of calcite that would precipitate when the groundwater is heated from 12°C to 99.9°C decreases by 45% when 2.4×10^{-3} moles/kg (228 mg/L) of phosphate are added to the water. The form in which the phosphate is added is an important aspect of the design. If it is added as trisodium phosphate (Na_3PO_4), the phosphate will also complex with free hydrogen ions in solution and raise the solution pH. The decrease in calcite solubility with increasing pH can easily surpass the increase in solubility due to phosphate complexation. Alternatively, if the phosphate is added as phosphoric acid (H_3PO_4), the decrease in pH caused by the added hydrogen ions will enhance the solubility of calcite.

TABLE 6-3.

INHIBITION OF CALCITE PRECIPITATION BY PHOSPHATE ADDITION

PO_4 added (moles/kg)	% Ca complexed with PO_4	Calcite precipitated (moles/kg)
1E-12	0	5.94E-04
6E-04	9	5.3E-04
1.2E-03	18	4.7E-04
1.8E-03	25	4.0E-04
2.4E-03	31	3.3E-04

Note: Composition of initial groundwater given in Table 6-2; pH = 6.71, temperature = 99.9C; closed system with respect to CO_2 gas.

The effect of adding acid on the solubility of calcite can be tested by simulating the addition of hydrochloric acid to the heated ATES groundwater. Table 6-4 shows the effect of adding small increments of acid to the heated groundwater, once again assuming that the system is closed to CO_2 gas exchange. Each incremental addition of acid changes the pH by only about 0.1 units, but the effect on calcite solubility and the amount of calcite that would precipitate is appreciable. If 1.5×10^{-3} moles/kg (36 mg/L) of HCl are added, the solution would be undersaturated with respect to calcite at 99.9°C. The main reason for this effect is the decreasing concentration of the carbonate species (CO_3^{2-}) compared with the other inorganic carbon species (H_2CO_3 and HCO_3^-) as the pH decreases. Because carbonate is a component of calcite, decreasing its concentration in solution increases the solubility of calcite.

TABLE 6-4.

INHIBITION OF CALCITE PRECIPITATION BY HCl ADDITION

HCl added (moles/kg)	pH	CO_3^{2-} (moles/kg)	Calcite precipitated (moles/kg)
0	6.71	1.5E-06	5.97E-04
5E-04	6.58	11E-06	3.7E-04
1.0E-03	6.49	8.6E-07	1.4E-04
1.5E-03	6.41	7.1E-07	−1.07E-4[a]

Note: Composition of initial groundwater given in Table 6-2; temperature = 99.9C; closed system with respect to CO_2 gas.

[a] Calcite undersaturated; negative value represents amount dissolved to reach equilibrium.

The method used at the St. Paul ATES site to eliminate calcite clogging problems was to install an ion exchange column upstream of the heating unit to remove calcium from the water and replace it with sodium.[4] Sodium will not form carbonate minerals at the temperatures and dissolved concentrations in the system. Appelo and Postma[5] simulated the injection and recovery of the sodium-rich, heated groundwater with the PHREEQM code. They were able to closely match the measured Na, Ca, and Mg concentrations in the recovered water, assuming that cation exchange was the dominant process affecting these constituents in the aquifer.

6.3. MODELING THE AERATION OF GROUNDWATER

Pumping groundwater to the land surface for use or remediation purposes may produce unwanted side effects due to equilibration of the water with gases in the atmosphere. The general process of solution/gas equilibration is discussed in Chapter 2. This section describes the development and use of a geochemical reaction model to estimate the impact of aeration on groundwater composition and the byproducts of gas exchange with the solution.

Groundwater is commonly in equilibrium with a CO_2 gas partial pressure much higher than the atmospheric value of 0.0003 atm. This occurs because oxidation of organic matter in the subsurface produces CO_2 as one of the byproducts. The high CO_2 content of groundwater lowers its pH and increases the solubility of minerals, particularly carbonate minerals. Under reducing conditions, also perhaps produced by the consumption of oxygen by organic matter, iron and manganese oxide and hydroxide minerals are also relatively soluble compared with oxidizing conditions. Aeration of groundwater causes a loss of carbon dioxide and an increase in oxygen as the solution equilibrates with atmospheric gas concentrations. The loss of CO_2 causes the pH of the water to rise, and the increase in O_2 causes the Eh to rise. These changes in pH and Eh may cause minerals to precipitate if their solubilities are exceeded.

Aeration can be simulated by equilibrating groundwater with atmospheric conditions using a geochemical model of the system and allowing reactive minerals to precipitate. Table 6-5 shows the groundwater environment as the initial condition. Environment I data are from a monitoring well located downgradient of a landfill. Conditions favor the reduced species of iron and manganese, and the oxyhydroxide minerals ferrihydrite [Fe(OH)$_3$] and manganite (MnOOH) are soluble (saturation indices less than 0). The equilibrium CO_2 partial pressure is 0.43 atm, which is over 1000 times greater than atmospheric conditions.

TABLE 6-5.

MODELING THE AERATION OF GROUNDWATER

Solution parameters[a]	Environment I Reducing, high metals & PCO$_2$	Environment II Aeration, equilb. with atmos. air	Environment III Aeration w/mineral pptn
Ca	4.18E-03	4.18E-03	9.03E-05
CO$_3$	2.98E-02	1.06E-02	3.70E-03
Fe	3.97E-04	3.97E-04	1.45E-08
Mn	2.10E-04	2.10E-04	7.40E-13
PCO$_2$ (atm)	0.43	0.0003 (fixed)	0.0003 (fixed)
pH	6.16	9.1	8.86
Eh (mV)	109	600 (fixed)	600 (fixed)
S.I. calcite	−0.45	2.27	0 (fixed)
S.I. ferrihydrite	−1.47	4.1	0 (fixed)
S.I. manganite	−8.1	9.5	0 (fixed)
Calcite pptd			4.09E-03
Ferrihydrite pptd			3.97E-04
Manganite pptd			2.06E-04

[a] Concentration units: moles/kg (constant concentration: Mg: 1.62E-3; Na: 2.5E-3; K: 2.26E-4; Cl: 2.09E-3).

Aeration is simulated by taking the original groundwater composition and fixing the CO_2 partial pressure at the atmospheric value and setting the redox potential at oxidizing conditions (Eh = 600 mV). The results are shown in Table 6-5 under Environment II. Because of the loss of CO_2 gas to the atmosphere, the pH has increased from 6.16 to a calculated value of 9.1. Under these conditions, calcite, ferrihydrite, and manganite are oversaturated as shown by positive saturation indices. Environment III shows the result of allowing these minerals to precipitate to equilibrium. Almost all of the iron and manganese are removed from the groundwater and a large amount of calcite precipitates. The majority of the precipitate is expected to be calcite because calcium and carbonate were the dominant dissolved species in the original groundwater. If the groundwater is pumped at a rate of 50 gpm, the amount of calcite that precipitates equals about 30 cc/minute (43 L/day). The amount of iron and manganese minerals predicted to precipitate are each less than 10% of the amount of calcite that might precipitate. Depending on how the water is used, the precipitation of these minerals may interfere with the piping or processing system.

To minimize the impact of mineral precipitation on water use or treatment, the loss of CO_2 could be controlled. Table 6-6 shows the results on mineral precipitation of simulating various CO_2 partial pressures between the atmospheric value and the groundwater value. The amount of calcite precipitated is proportional to the CO_2 partial pressure, which also impacts the pH of the water. The higher the CO_2 pressure, the lower the pH and the smaller amount of calcite precipitates. By maintaining the CO_2 partial pressure at 0.1 atm, the amount of calcite that precipitates can be reduced by a factor of ten from the amount calculated for atmospheric air conditions. The impact of CO_2 on the amounts of ferrihydrite and manganite that precipitate is not as significant because under oxidizing conditions these minerals do not become appreciably soluble until the pH is lower than CO_2 gas can produce. As shown in Table 6-6, it is not until the CO_2 partial pressure is 0.5 atm and the pH is 6.26 that manganite solubility begins to increase. Even at this P_{CO_2}/pH, there is no increase in ferrihydrite solubility.

TABLE 6-6.

MINIMIZING MINERAL PRECIPITATION FROM AERATED GROUNDWATER BY CO_2 ADDITION

Solution parameters	Aerated groundwater[a]	Controlled CO_2 partial pressure (atm)			
		0.001	0.01	0.1	0.5
pH	8.76	8.29	7.45	6.71	6.26
Calcite pptd[b]	4.09E-03	3.90E-03	3.03E-03	3.54E-04	0
Ferrihydrite pptd	3.97E-04	3.97E-04	3.97E-04	3.97E-04	3.97E-04
Manganite pptd	2.06E-04	2.06E-04	2.06E-04	2.06E-04	1.78E-04

[a] CO_2 partial pressure equals atmospheric value of 0.0003 atm.
[b] Moles/kg.

To eliminate the precipitation of calcite and the metal oxyhydroxides, an acid such as HCl may be added to the groundwater as it equilibrates with the atmosphere. The results of this simulation are shown in Table 6-7. In this case the CO_2 partial pressure is set at atmospheric conditions and the Eh is set at 600 mV. The decrease in the pH is due to the addition of the acid. As the pH decreases a few tenths of a unit, the amount of calcite that precipitates decreases dramatically. At a given pH, the amount of calcite that precipitates is lower when HCl is added compared with minimizing precipitation by maintaining elevated CO_2 levels. The reason for this is that high CO_2 levels increase the overall carbonate concentration of the solution, requiring more calcite to precipitate to reach equilibrium compared with a water in equilibrium with a lower P_{CO_2} value. When 0.012 moles/kg of HCl has been added to the groundwater in equilibrium with an atmospheric P_{CO2} level, the pH is lowered to 7.11 and calcite no longer precipitates. However, the iron and manganese mineral precipitation has not been affected by lowering the pH to 7.11. Adding a very small amount (0.001 mole/kg) of additional HCl lowers the pH to a value close to 4. At this pH, manganite is soluble, even under oxidizing conditions, and ferrihydrite is slightly more soluble. An additional small incremental increase in acid would lower the pH to less than 4 and eliminate the precipitation of ferric hydroxide; however, the water would be fairly corrosive and may not be suitable for its intended use. An alternative to pH control to minimize the precipitation of the metals would be to allow the metals to precipitate and filter the particulates from the solution.

TABLE 6-7.

MINIMIZING MINERAL PRECIPITATION FROM AERATED GROUNDWATER BY HCl ADDITION[a]

Solution parameters	Aerated groundwater[a]	Amount of HCl added (moles/kg)			
		0.005	0.01	0.012	0.013
pH	8.76	8.19	7.97	7.11	4.04
Calcite pptd[b]	4.09E-03	3.02E-03	7.26E-04	0	0
Ferrihydrite pptd	3.97E-04	3.97E-04	3.97E-04	3.97E-04	3.21E-04
Manganite pptd	2.06E-04	2.06E-04	2.06E-04	2.06E-04	0

[a] CO_2 partial pressure equals atmospheric value of 0.0003 atm.
[b] Moles/kg.

Mineral precipitation in response to the aeration of groundwater is the major maintenance cost of air-stripping towers used to remove volatile organic contaminants from groundwater. Geochemical modeling can be used with the site-specific groundwater conditions to estimate whether mineral precipitation will be a problem at a particular site and to calculate the amount of reagent required to minimize the amount of precipitate formed.

REFERENCES

1. Chapelle, F.H. and Lovley, D.R., Competitive exclusion of sulfate reduction by Fe(III)-reducing bacteria: a mechanism for producing discrete zone of high-iron ground water, *Ground Water*, 30, 29–36, 1992.
2. Bennett, P.C., Siegel, D.E., Baedecker, M.J., and Hult, M.F., Crude oil in a shallow sand and gravel aquifer. I. Hydrogeology and inorganic geochemistry, *Appl. Geochem.*, 8, 529–549, 1993.
3. Holm, T.R., Eisenreich, S.J., Rosenberg, H.L., and Holm, N.P., Groundwater geochemistry of short-term aquifer thermal energy storage test cycles, *Water Resour. Res.*, 23, 1005–1019, 1987.
4. Perlinger, J.A., Almendinger, J.E., Urban, N.R., and Eisenreich, S.J., Groundwater geochemistry of aquifer thermal energy storage: long-term test cycle, *Water Resour. Res.*, 23, 2215–2226, 1987.
5. Appelo, C.A.J. and Postma, D., *Geochemistry, Groundwater and Pollution*, A.A. Balkema, Rotterdam, 1994.

7

GEOCHEMISTRY OF CONTAMINANT MOBILITY

Contamination in the subsurface may be thought of as the presence of any chemical compound at a level and in a form that may cause damage to human or environmental receptors. Many naturally occurring trace elements (e.g., Cu, Cr, Se, and Zn) are considered essential for human nutrition,[1] but may be toxic at high concentrations. The form of the compound (solid, gas, and/or solution phase) affects the manner of uptake of the compound. Solid phase and dissolved contaminants may be ingested by the receptor or be assimilated through the skin, and gaseous contaminants may be inhaled or also assimilated through dermal contact. Water/rock/gas interactions distribute contaminants between the phases, which impacts the mobility of the contaminant and the opportunity for exposure. In general, the solution and gas phases are more mobile and can more widely disburse contaminants than are the solid phases. Geochemical processes can control and, perhaps, limit contaminant concentrations in the various phases, thereby directly impacting exposure levels.

In this chapter the concept of contaminant mobility will be considered from the standpoints of the impact of an introduced contaminant on geochemical equilibrium (Figure 7-1A) and the potential for release of a contaminant by a change to the environment (Figure 7-1B). In the first case a new compound contaminates the environment and geochemical reactions affect its mobility in the subsurface. The contaminant (C) will distribute itself between the solution phase (C_{aq}), the gas phase (C_{gas}) if it is volatile, the surface of solids (C_{ads}), and perhaps as a component of a mineral (C_{solid}) if the mineral precipitates. In the second case, a relatively immobile contaminant initially present in the system is mobilized by changing environmental conditions, for example, raising or lowering the pH. In each case the mobility of the contaminant will be affected by the concentration it can reach in the more mobile solution and gas phases. Concentrations in the mobile phases will be limited by mineral and gas solubilities and adsorption/desorption processes on the surfaces of aquifer solids.

7.1. CONTAMINANT-INDUCED DISEQUILIBRIUM

Contaminants entering an environment, whether natural or man-induced, are a disturbance to that environment. According to Le Chatlier's Rule, the geochemical system will respond

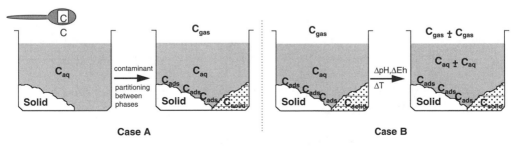

Figure 7-1 Contamination scenarios.

by minimizing the effects of the disturbance. The response will generally be to lower contaminant concentration in the phase in which the contaminant is introduced. For example, if equilibrium exists between a dissolved metal (M^+) in groundwater and adsorption sites (e.g., $\equiv FeOH$) on the surface of the aquifer solids according to the following reaction:

$$\equiv FeOH + M^+ \leftrightarrow \equiv FeOH + H^+ \tag{7-1}$$

then increasing the solution concentration level of the metal will drive the reaction to the right, causing some of the added metal to adsorb. This will establish a new equilibrium distribution of the metal between the solution and solid surface. The overall result is that some of the added contaminant metal is removed from the mobile solution phase and attached to the immobile solid phase. The contaminant metal will compete with the other adsorbed metals and displace an amount such that the final concentrations are in equilibrium according to the adsorption affinity of the surface for each of the competing metals. The result will be a decrease in solution concentration for the added contaminant and an increase in solution concentration for the other metals in the system. The system adjusts concentration levels in its various phases to lower the concentration of the added dissolved metal. The solid phase concentration of the added metal has increased and displaced other adsorbed metals, increasing their solution concentration.

Disequilibrium conditions may also enhance metal mobility. If uncontaminated groundwater enters a zone in an aquifer where adsorbed metal concentrations are locally elevated, then disequilibrium will drive Reaction 7-1 to the left, resulting in desorption of the metal. The dissolved concentration of the metal will increase until equilibrium is reestablished at a higher dissolved metal level. These two examples of disequilibrium effects on metal mobility show that geochemical reactions tend to lower contaminant concentrations in whatever phase contamination is added to the system. Adding contaminated water to clean soil will lower contaminant levels in the water as the levels rise in the soil. Adding contaminated soil to clean water will contaminate the water as soil quality improves. In each case the reaction progresses in the direction that will eliminate the disequilibrium imposed by contamination.

In addition to the adsorption/desorption response to disequilibrium, the system may react by dissolving or precipitating a mineral containing the contaminant or dissolving or exsolving a gas. If the introduction of a contaminant produces oversaturation with respect to a reactive mineral, then the mineral should precipitate removing enough of the contaminant and the other components of the mineral to reestablish equilibrium. If a gas such as H_2S is produced by contamination in an aquifer, then the gas will be lost from solution to the vapor phase until equilibrium concentrations are attained in the solution and gas phases in accordance with the solubility of the gas. Alternatively, if volatile contaminants are present in the unsaturated zone, they will dissolve to their equilibrium solution concentration in the groundwater. This will lower concentrations in the vapor phase, but produce contamination in the groundwater. The solubility of the gas in solution will be a function of its partial pressure in the vapor phase, its partitioning between the solution and gas phase (Henry's Law), and speciation reactions in the solution (see Section 2.3).

7.2. SOLUBILITY LIMITS ON CONTAMINANT CONCENTRATION IN GROUNDWATER

The solubility of a mineral containing a contaminant may limit solution concentration of the contaminant. The equilibrium concentration may or may not be below the maximum contaminant level (MCL). For example, consider the case where barite ($BaSO_4$) is a reactive mineral in equilibrium with groundwater. The equilibrium reaction can be written as follows:

$$BaSO_4 \leftrightarrow Ba^{2+} + SO_4^{2-} \quad K(25°C) = 10^{-9.98} \tag{7-2}$$

The drinking water standard for barium is 2 mg/L (1.46×10^{-5} moles/L). The dissolved sulfate concentration in equilibrium with barite at a dissolved barium concentration of 2 mg/L can be approximated from the equilibrium relationship as follows:

$$\left(SO_4^{2-}\right)\left(Ba^{2+}\right) = 10^{-9.98} \tag{7-3}$$

$$\left(SO_4^{2-}\right) = 10^{-9.98}/\left(Ba^{2+}\right) = 10^{-9.98}/10^{-4.48} \tag{7-4}$$

$$\left(SO_4^{2-}\right) = 10^{-5.14} \text{ moles/L} = 0.7 \text{ mg/L} \tag{7-5}$$

(This is an approximation because the effects of complexation and activity corrections have not been considered. These effects will tend to increase the solubility of barite and the equilibrium barium and sulfate concentrations.) As long as the dissolved sulfate concentration can be maintained at about 0.7 mg/L or higher, the dissolved barium concentration should not exceed 2 mg/L because the precipitation of barite will remove barium to this level or lower. Therefore, if a solution containing barium at a level above the MCL enters this system, barite will limit the solution concentration to an acceptable level. Alternatively, if the sulfate concentration is less than about 0.7 mg/L and barite is present in the system, then the dissolved barium concentration will be greater than 2 mg/L at equilibrium with barite. In this case any additional dissolved barium added to the system will remain mobile, and barite will not limit barium concentration to an acceptable level. For a multicomponent mineral such as barite, the concentration of each component in solution must be considered in evaluating whether the mineral will limit the concentration of the contaminant to an acceptable level.

Sulfate is an example of a potential contaminant that typically forms fairly soluble minerals, except for barite, that do not limit sulfate concentration to a low level. In most natural groundwaters the sources of sulfate exceed the sources of barium; consequently, if barite forms, the mineral limits the barium concentration and not the sulfate concentration. As the sulfate concentration increases in the groundwater, it may reach equilibrium with gypsum ($CaSO_4 \cdot H_2O$), which often provides an upper limit on dissolved sulfate concentration in aquifer systems. However, because gypsum is a relatively soluble mineral, the groundwater sulfate concentration in equilibrium with gypsum may be in the range of thousands of milligrams per liter, depending upon the degree of complexation of the sulfate in solution. Where gypsum is present in an aquifer, it generally does not limit sulfate concentration to an acceptable level. In many cases the reverse is true; gypsum is a source of sulfate at an unacceptable level for drinking water.

When considering anticipated contaminant solution concentrations, it is best to think in terms of the solubility of minerals containing the contaminant. It is not appropriate to think in terms of the "solubility of an element," as in "iron is soluble under reducing conditions and insoluble under oxidizing conditions." It is not the iron that is soluble or insoluble; it is the minerals containing iron. Dissolved iron may be high or low under oxidizing or reducing conditions, depending on other factors such as pH and the presence of other dissolved species (such as sulfide) that might form a mineral with the iron.

7.3. MINERAL EQUILIBRIA OF CONTAMINANTS

In considering the potential for mineral equilibrium control on contaminant solution concentration, it is necessary to evaluate both the solubility of the mineral and the concentration of other dissolved components of the mineral. As described above for sulphate and gypsum, highly soluble minerals may not limit solution concentrations to acceptable levels. For multicomponent minerals, if the solution concentration of the noncontaminant component is low this may allow the dissolved concentration of the contaminant component to be unacceptably high even if the mineral is relatively insoluble (for example barite). For contaminants that form oxide and hydroxide minerals (e.g., MnO_2 and $Cr(OH)_3$), the concentration of the oxide and hydroxide components of the minerals is considered unlimited in groundwater and would not impact the ability of the mineral to remove the contaminant from solution as the mineral phase. However, the solubility of the mineral is dependent on other solution parameters such as pH, Eh, and ionic strength. The following provides general guidelines on the types of minerals that form in aquifer environments and may limit the solution concentrations of contaminants that are components of these minerals. Appendix A contains a comprehensive list of reactive minerals.

Simple Metals — Ba, Cd, Pb, Ni, and Zn

For the purposes of contamination evaluation, simple metals are those that are not redox sensitive although other components of minerals containing these metals may be affected by the redox potential of the system. Under oxidizing conditions, Ba, Cd, Pb, Ni, and Zn can form carbonate ($BaCO_3$, $CdCO_3$, $PbCO_3$), sulfate ($BaSO_4$), oxide/hydroxide [$Pb(OH)_2$, $NiFe_2O_4$, $Ni(OH)_2$, $ZnFe_2O_4$], and phosphate [$Cd_3(PO_4)_2$] minerals. Under reducing conditions with sufficient sulfide present, Cd, Ni and Pb may form sulfide minerals that are generally very insoluble as long as reducing conditions are maintained.

Simple Nonmetals — PO_4 and F

The concentration of fluoride in groundwater is typically limited by formation of the minerals fluorite (CaF_2) and fluorapatite [$Ca_5(PO_4)_3F$]. In addition to fluorapatite, phosphate may be limited by the formation of apatite ($Ca_3PO_4)_2$), variscite ($AlPO_4 \cdot 2H_2O$), strengite ($FePO_4 \cdot 2H_2O$) and $MnHPO_4$. In groundwater, phosphate may occur as polyphosphate species (such as $H_4P_2O_7$) from detergents and fertilizers and as organic compounds both of which are less reactive and more mobile than orthophosphate (PO_4^{3-}).

Redox-Sensitive Metals — Fe, Mn, Cr, Hg, and Cu

Under oxidizing conditions at pH values greater than about 5.5, these metals, which can occur in more than one valence state, form relatively insoluble minerals. Total iron solution concentration is oftentimes limited by ferrihydrite ($Fe(OH)_3$) and manganese concentration is limited by MnO_2. Chromium forms an hydroxide mineral ($Cr(OH)_3$) that has low solubility under these conditions and Hg may be found in an iodide mineral (Hg_2I_2) or as elemental mercury ($Hg(l)$). Copper can form a cupric ferrite mineral that has the formula $CuFe_2O_4$.

Under reducing conditions with sulfide present, Fe and Hg form insoluble sulfide minerals. Manganese may be present in the solid phase as manganite ($MnOOH$), Mn_2O_3, or rhodochrosite ($MnCO_3$). The chromium hydroxide mineral $Cr(OH)_3$ may still be the stable phase or, if sufficient iron is present, an iron-chrome mineral ($FeCr_2O_4$) may be stable. Copper may form the cuprous ferrite mineral $Cu_2Fe_2O_4$).

Redox-Sensitive Nonmetals — As and Se

Arsenic and selenium form fairly soluble arsenate [As(V)] and selenite [Se(IV)] minerals under oxidizing conditions in combination with iron ($FeAsO_4$, $Fe_2[OH]_4[SeO_3]$), lead ($Pb_3[AsO_4]_2$), and manganese ($Mn_3[AsO_4]_2$). Under reducing conditions, elemental selenium may limit solution concentrations, as well as the iron selenide $FeSe_2$. Arsenic forms very insoluble sulfide minerals such as orpiment (As_2S_3) under reducing conditions.

7.4. ADSORPTION/DESORPTION IMPACT ON CONTAMINANT MIGRATION

The impact on contaminant migration of partitioning of a contaminant between the groundwater and the surfaces of the aquifer solids can be most easily demonstrated in the case of the simple distribution coefficient. The distribution coefficient (Kd, L/kg) is the ratio of the concentration of a contaminant adsorbed on the surfaces of the solid phase (C_s, mg/kg solid) to the dissolved concentration (C_{aq}, mg/L) of the contaminant.

$$Kd = C_s/C_{aq} \qquad (7\text{-}6)$$

Figure 7-2 shows the impact of adding a contaminant to a water/rock system when the Kd for the contaminant is equal to 2 L/kg. In this case the concentration adsorbed equals twice the concentration in solution; however, the adsorbed mass of the contaminant is greater than two times the mass in solution because of the units used for Kd. C_s is in terms of milligrams of contaminant per kilogram of aquifer solid, whereas C_{aq} is in terms of milligrams of contaminant per liter of water. The amount of aquifer solid in contact with a liter of water under saturated conditions is the bulk density of the solid divided by the porosity of the aquifer, p_b/η. For example, if the bulk density is 2 g/cc and the porosity is 0.25, the mass of the solid phase in contact with one liter of groundwater is 8 kg. Therefore, if the Kd is 2 and the equilibrium solution concentration is 1 mg/L for a contaminant, then the solid phase concentration is 2 mg/kg and the mass of contaminant on the solid is 16 mg. For these conditions the mass of the contaminant on the solid is 16 times the mass in solution when considering a unit volume of the aquifer. Figure 7-2 shows the adsorbed mass for a contaminant with a Kd of 2 at several solution concentrations.

C_{aq} (mg/L)	C_{ads} (mg/kg)	Aqueous Mass per liter water (mg)	Adsorbed Mass per liter water (mg)
0.5	1	0.5	8
1	2	1	16
1.5	3	1.5	24
2	4	2	32

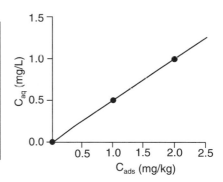

Figure 7-2   Linear contaminant adsorption (Kd = 2).

The total mass of the contaminant in the system per liter of groundwater ($M_{C,T}$) is the sum of the mass dissolved in solution and the mass adsorbed onto the solid.

$$M_{C,T} = \left(C_{aq} \times 1 \text{ liter}\right) + p_b/\eta \left(C_s \times 1 \text{ liter}\right) \qquad (7\text{-}7)$$

The total concentration (C_T, in mg/L) of the contaminant in an aquifer volume occupied by one liter of water is as follows:

$$C_T = C_{aq} + \left(p_b/\eta\right)C_s \qquad (7\text{-}8)$$

The definition of Kd can be used to determine C_{aq} and C_s in terms of C_T as follows:

$$C_s = \frac{C_T Kd}{1 + (\rho_b/\eta)Kd} \tag{7-9}$$

$$C_{aq} = C_T - (\rho_b/\eta)C_s \tag{7-10}$$

Figure 7-3 shows the effect on concentrations of partitioning between the solution and solid phase as a contaminant moves through an aquifer interacting with a solid phase that initially has none of the contaminant adsorbed on its surfaces. In Box 1, 17 milligrams of the contaminant is added to a liter of groundwater. The Kd of the contaminant is 2; therefore the resulting solution concentration is 1 mg/L and the solid phase concentration is 2 mg/kg. The mobile, dissolved contaminant is allowed to migrate to an adjacent position in the aquifer (Box 2), occupying one liter of pore space. In this case the 1 mg of total mass of mobile contaminant partitions again between the initially clean aquifer surfaces and the groundwater resulting in 0.059 mg/L in solution and 0.118 mg/kg on the solid. As can be seen from Figure 7-3, the dissolved concentration of the contaminant decreases by a factor of 17 [$= 1 + (\rho_b/\eta)Kd$] for each step as adsorption retards the movement of the contaminant relative to the flow of water. As the water moves downgradient, the majority of the contaminant is left behind adsorbed onto the solid.

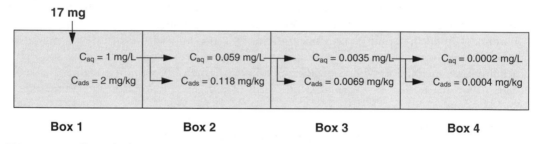

Figure 7-3   Effect of adsorption on contaminant mobility (Kd = 2 L/kg; ρ_b = 2 g/cc; η = 0.25).

Figure 7-4 shows the effect of desorption on the release of a contaminant initially present adsorbed onto the aquifer surfaces. The initial condition is the same for the previous case in which the starting total mass of contaminant is 17 mg divided between 1 mg/L dissolved in the groundwater and 2 mg/kg adsorbed onto the solid. At a bulk density of 2 g/cc and a porosity of 0.25, the total mass of contaminant adsorbed is 16 mg. In Step 2, the contaminated groundwater flows downgradient and is replaced by fresh groundwater without any initial contaminant. To reach equilibrium between the fresh groundwater and the contaminant on the aquifer surface, some of the contaminant desorbs until the ratio of adsorbed to dissolved concentrations equals the Kd for this site, which is 2. The result is that the initially uncontaminated water now has 0.94 mg/L of the contaminant in solution. The surfaces contributed this contaminant to solution slightly decreasing the adsorbed concentration to 1.88 mg/kg. As shown by the additional steps in Figure 7-4, the solid phase acts as a continuing source of the contaminant as additional fresh groundwater passes through the zone of contamination. For these conditions the water in the zone of contamination would have to be replaced 50 times by freshwater to lower the solution concentration from 1 to 0.05 mg/L.

The simple Kd approach to evaluating the impact of adsorption/desorption on contaminant mobility is not appropriate in cases where the system parameters (e.g., pH, Eh, concentration of major and competing ions, and sorbents) change along the flow path or

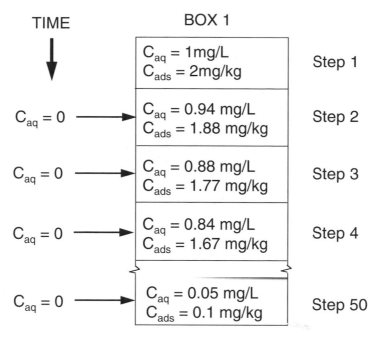

Figure 7-4   Effect of desorption on contaminant concentration (Kd = 2 L/kg; ρ_b = 2 g/cc; η = 0.25).

when the concentration of the contaminant approaches the adsorption capacity of the sorbents. In these cases more sophisticated methods of evaluating adsorption/desorption must be employed (see Section 3.1). However, the general effects of adsorption retarding the migration of contaminants and desorption providing a continuing source of contamination applies in all cases. The impact of adsorption/desorption on aquifer restoration is discussed in Chapter 8.

7.5. CONTAMINANT MIGRATION SUMMARY

The combined impact of adsorption/desorption and mineral equilibria on contaminant mobility is shown schematically in Figure 7-5. This graph equates the dissolved concentration of a contaminant to the concentration of the contaminant associated with the solid phases, either adsorbed or present as a component of a mineral. At low dissolved concentrations, adsorption processes dominate the partioning between the solution and solid phases. As additional contamination is added to the system, more of the contaminant is adsorbed, but the resulting solution concentration also increases along the sloping line. The equilibrium point of the system moves up the slope of the line. The *slope* of the line in the zone of adsorption is dependent on the affinity of the solid for the contaminant. If the solid has a strong affinity (high Kd) for the contaminant, the slope will not be very steep and additional contaminant added to the system will preferentially be adsorbed and removed from solution compared with a contaminant with a low Kd. For a contaminant with a low Kd, the slope of the line will be steep and the concentration of the contaminant in solution will be high relative to that on the solid.

As the aqueous concentration increases, the solution may become saturated with respect to a mineral phase containing the contaminant. Assuming that the other components of the mineral are present in adequate supply, the solution concentration of the contaminant will not increase beyond the saturation concentration, while the amount of the contaminant present as the solid phase increases as the mineral precipitates. This is the zone of mineral-

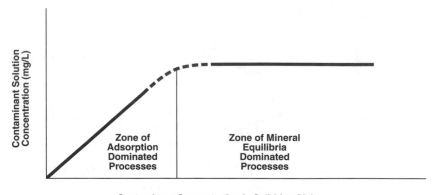

Figure 7-5 Adsorption and mineral equilibrium effects on contaminant solution concentration/mobility.

dominated processes on Figure 7-5. Mineral precipitation provides an additional mechanism for accumulation of the contaminant in the aquifer.

The slope of the adsorption/desorption line and its intersect with the saturation concentration line are highly dependent on solution parameters (e.g., pH, Eh, temperature, composition) and the types of solid phases present. At higher solution concentrations, the slope of the adsorption/desorption line may begin to curve as the contaminant fills the available sites for adsorption. The point at which mineral saturation is reached may vary by orders of magnitude for a contaminant depending on the other constituents in solution and the pH and Eh of the solution. For example, the solubility of ferrihydrite at a pH of 7 can increase by a factor of almost 3000 if the redox potential is lowered from +200 mV to 0 mV. Site-specific conditions must be considered in evaluating the importance of adsorption/desorption and mineral equilibrium on contaminant mobility in a particular aquifer. In addition, nonequilibrium conditions may prevail at a site where reaction rates are slow and mineral equilibrium, in particular, is not attained during the time frame that contaminant disequilibrium has existed in the system.

REFERENCE

1. Campen, D.R.V., in *Micronutrients in Agriculture*, Mortvedt, J.J., Cox, F.R., Shuman, L.M., and Welch, R.M., Eds., Soil Science Society of America, Madison, WI, 1991, pp. 663-701.

8 GEOCHEMISTRY OF RESTORATION

Aquifer restoration includes the natural processes and applied techniques that lower contaminant concentrations in the groundwater and/or solid phase to below a cleanup level. Physical processes occurring in an aquifer such as dispersion, diffusion, and mixing will naturally decrease contaminant concentration levels in groundwater. Both natural and enhanced biodegradation will lower solution and solid phase concentrations of organic contaminants as organisms use the contaminant substances as a source of food, thereby converting the contaminant to another form of organic carbon or to inorganic carbon. This chapter focusses on the geochemical processes active in an aquifer that enhance or inhibit restoration. The most direct processes are mineral precipitation/dissolution and adsorption/desorption; however, the effectiveness of these processes is a function of many other geochemical processes such as aqueous complexation, oxidation/reduction, and neutralization.

Because an aquifer consists of both solution and solid phases that interact, it is necessary to consider the entire system of all reactive phases when evaluating aquifer restoration. In addition, restoration may require not just cleanup of the groundwater but also reduction of contaminant levels in the solid phase if the solids themselves represent a potential threat to human health or the environment. In this case the aquifer solids need to be considered not just because they provide a continuing source of groundwater contamination, but because they themselves are a contaminated medium included in the overall risk calculations for the site. In this chapter the geochemistry of aquifer restoration will be considered for both the solution and solid phases in the subsurface. Natural geochemical processes that enhance restoration will be described. General geochemical principles to consider in designing and implementing applied techniques are discussed, as well as the potential impacts of geochemistry on various applied restoration methods. Simulating restoration with geochemical models will also be described.

8.1. NATURAL RESTORATION

Natural, or intrinsic, restoration includes all those physical, biological, and geochemical processes active in the system that lower concentrations in the contaminated media to below a level of concern. The physical processes of dispersion, diffusion, and mixing generally dilute the dissolved contaminant and spread the lower concentration level of contamination over a larger volume of groundwater. The physical process of volatilization occurs at the water table and in the unsaturated zone. It transfers the contaminant from the solution to the gas phase or vice versa, depending on the level of saturation of the gaseous phase of the contaminant in the groundwater. Biodegradation lowers solution concentrations of reactive organic contaminants principally by oxidation-reduction reactions. The product(s) of these bacterially mediated reactions may be more or less toxic than the original contaminant.

Geochemical processes lower contaminant concentrations in groundwater by adsorbing the contaminant onto the surfaces of the minerals and organic solids naturally present in the aquifer and by incorporating the contaminant as part of a mineral or amorphous solid that is precipitating in the aquifer environment. Whether by adsorption or mineral precip-itation, the overall effect is to lower contaminant concentration in the groundwater and increase the concentration level in the solid phase. These are the same reactions that impacted the migration of contaminants described in Chapter 7. If the natural system is capable of immobilizing contaminants emanating from a source area and lowering dissolved concentrations to below a level of concern, then the system has successfully remediated itself, assuming that the concentration of the contaminant produced in the solid phase has not exceeded a level of concern for this phase.

As in contaminant mobility evaluations, the potential effectiveness of natural restoration at a contaminated site is a function of the disequilibrium in the system produced by the presence of the contaminant and the reactivity of the system to eliminate the disequilibrium over an acceptably short period. For example, acidic leachate containing high dissolved metal concentrations will react with carbonate minerals in the subsurface. If the acidity of the leachate does not exceed the acid neutralizing capacity of the aquifer, commonly provided by the carbonate minerals, the aquifer system may neutralize the leachate and lower the metals concentrations to low values in response to adsorption and precipitation of minerals containing the metals. In this case there is initially a strong disequilibrium between the acid leachate and the carbonate-containing solids in the aquifer. Because the carbonate minerals react relatively quickly to the acidic conditions, they have the potential to remove at least some of the contaminants from solution and raise the pH of the leachate. The ability of the carbonate minerals to neutralize the solution may be compromised by the precipitation of metal oxyhydroxide minerals on the surface of the carbonates, thereby armoring the carbonate minerals from further reaction with the acidic solution. For this reason the total carbonate mineral content of the aquifer may not be available to participate in neutralization reactions. In this neutralization scenario, reactive minerals must be present to respond to the disequilibrium produced by the acid, the capacity of the natural system must be sufficient to neutralize the added acidity, and the reactive minerals must be in contact with the acidic solution to achieve neutralization.

Redox interfaces along groundwater flow paths are well known locations for the concentration of metals and other elements as a result of natural processes.[1,2] A redox interface typically occurs where the redox potential shifts to favor one redox state of an element over a more oxidized or reduced form of the element. The redox interface between Fe(III) and Fe(II) is most commonly recognized because a change in color of the aquifer is associated with this interface. Under more oxidizing conditions, typically upgradient of the iron redox interface, the stable iron solid is an Fe(III) mineral, normally ferrihydrite or goethite. These minerals color the aquifer in shades of yellow, brown, and red. Downgra-dient of this redox interface under more reducing conditions, the Fe(III) minerals are soluble and either do not form or do not persist. The color of the aquifer under these conditions will reflect that of the primary minerals and may be light in color if quartz, feldspars, or calcite are the dominant minerals, but most likely the darker minerals (biotite, amphibole, and pyroxenes) will produce a gray or black color. This is particularly true if sulfide is present and the dark mineral mackinawite (FeS) has precipitated at the iron-reducing interface. Iron oxidizing interfaces also occur where reducing water that favors Fe(II) minerals is oxidized and Fe(III) minerals precipitate in the aquifer. This situation can occur naturally when anoxic groundwater mixes with oxygenated water or is exposed to atmo-spheric air in a well[3] and when groundwater contaminated with organic compounds from, for instance, a landfill or leaking underground storage tank mixes downgradient with uncontaminated, oxygenated water.[4,5]

At the iron-reducing interface several elements such as Se, As, Mo, and U may be concentrated because of the lower solubility of minerals containing these elements under reducing conditions than under oxidizing conditions.[6] The typical uranium mineral present at this interface is uraninite (UO_2). The elevated concentration of uranium in the solid phase at this interface is taken advantage of by locating and mining these occurrences, which are

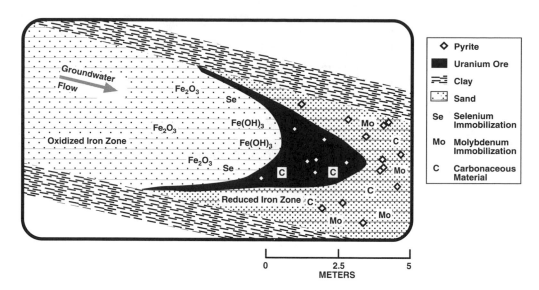

Figure 8-1   Uranium roll-front mineralization.

called roll-front uranium deposits because of their distinctive pattern. Figure 8-1 shows a cross section of a roll-front deposit with the direction of groundwater flow from the oxidizing to reduced zones of the aquifer. These deposits occur in aquifers straddled by confining units. The shape of the roll-front is due to the relatively high permeability in the mid-region of the aquifer compared with the lower permeability as the confining units with their higher clay content are approached. The majority of the flow is down the middle of the aquifer; therefore oxidizing conditions have moved farther downgradient in the center of the formation comprising the aquifer than at its top and bottom.

If the roll-front uranium deposit is present at a depth greater than a few tens of feet below the land surface, it is typically more economical to extract the uranium using an *in situ* leaching technique. Because the uranium minerals are relatively soluble under oxidizing conditions, especially when dissolved carbonate is present to complex with U(VI), the *in situ* technique normally involves simply adding oxygen and carbonate to groundwater and pumping it through the roll front. The injection and extraction system is tuned to minimize loss of leaching solution from the zone of the roll front, and perimeter wells are used to monitor water quality up- and downgradient of the roll front and in adjacent aquifers, if present.

At the termination of leaching, when the majority of the uranium has been removed from the roll front, restoration may be required to eliminate high residual levels of uranium in the groundwater. Prior to mining activities, uranium was relatively immobile at the roll-front redox interface due to the low solubility of uraninite under reducing conditions. Unless the reducing capacity of the aquifer has been eliminated, it is likely that the natural aquifer conditions still present downgradient from the zone of *in situ* leaching will tend to immobilize dissolved uranium because it will reprecipitate as uraninite. The potential effectiveness of this natural restoration process was evaluated in the laboratory for an *in situ* leach uranium mining site in south Texas. Leaching solution with a high dissolved uranium concentration (52 mg/L) was pumped through a column of aquifer material collected downgradient from a roll-front deposit.[7] As shown in Figure 8-2, the effluent concentration from the column was always well below the influent value, and after a few pore volumes of solution had passed through the column, the effluent concentration stabilized at a low value less than 1 mg/L. It is apparent at this site in Texas that the natural system had sufficient reducing capacity to reduce U(VI) in the leaching solution to U(IV) and apparently immobilize it as the mineral uraninite.

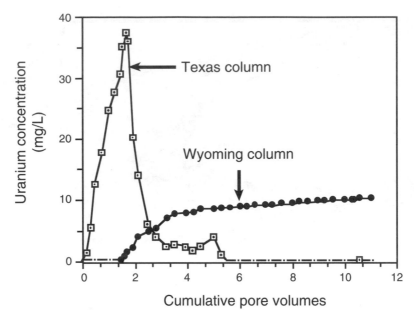

Figure 8-2   Uranium concentration in effluent from Texas and Wyoming natural restoration column experiments.

The ability of the natural system to remove uranium from groundwater at the cessation of *in situ* leach mining was also evaluated at a mine site in Wyoming. Similar column experiments were run as those for the Texas site; however, Figure 8-2 shows that the response was quite different. For the Wyoming site the effluent uranium concentration started low and then increased to the influent value of 14 mg/L. This shows that the Wyoming aquifer material did not have sufficient reducing capacity or reactivity under the chosen experimental conditions to immobilize uranium in residual leaching solution at a starting concentration of 14 mg/L. The reason for the difference in response of the aquifer materials is believed to be due to the amount of pyrite available as a reducing agent. The south Texas aquifer material had an appreciable amount of pyrite (4 wt. %), whereas only about 0.05 wt. % pyrite was detected in the Wyoming aquifer samples. The presence of the redox interface and the concentration of uranium at the Wyoming site may have been due to past conditions, such as the movement of H_2S gas through the aquifer, that no longer exist and did not provide a long-term reducing capacity to the system. This difference in response of the Wyoming and Texas aquifer materials shows the necessity of evaluating individual sites on their ability to immobilize the contaminants of interest by natural processes.

Neutralization reactions for acidic leachate and reducing fronts for certain metals are only two types of geochemical barriers that can perhaps contribute to natural restoration. Oxidizing, acidic, adsorption, and thermodynamic barriers have also been described.[1] Combination barriers, such as neutralization/adsorption barriers where iron and manganese oxides precipitate and adsorb other dissolved cations and anions onto solid surfaces, may also occur in the natural environment. The natural system will generally have some ability to reduce contaminant levels in groundwater. If the contaminant flux does not exceed the capacity of the natural system to respond and the natural system can immobilize contaminants below a level of concern, then natural restoration has been effective. If the contaminant flux is too great for the natural system or the water/rock interactions are not fast enough to immobilize the contaminants, then an active restoration method may be required to achieve cleanup levels or to bring contaminant concentrations down to a level where the natural system can be effective.

8.2. GEOCHEMICAL PRINCIPLES OF AQUIFER RESTORATION

In the design and implementation of an active restoration method, several general geochemical principles should be kept in mind so that the natural system can be used to enhance restoration as much as possible and not inhibit the removal or immobilization of the contaminants. These principles consist of starting from an adequate understanding of the natural and contaminated geochemical systems, working toward stable chemical equilibria, and identifying and avoiding unwanted side effects.

Understanding the Geochemical System

Because the geochemical system is reactive and will respond to an applied restoration technique such as pump-and-treat or injection of treatment chemicals, it is necessary to understand the important water/rock interactions at a site that are affecting concentrations in the groundwater and solid phases. This requires complete and accurate data on the groundwater composition (temperature, pH, Eh, dissolved oxygen, major ions, and trace constituents of interest) and on the distribution and concentration of reactive solids along the flow path. The selection of solids to measure in the system will be guided by the type of contaminant and potential remedial measures. For example, if organic contamination is present, then the concentration of organic carbon in the solid phase is necessary to calculate the distribution coefficient of the contaminants. The distribution coefficient can then be used to evaluate mobility and the effectiveness of a pump-and-treat removal system. If the contaminating solution is acidic, then data on the acid neutralizing capacity of the aquifer is necessary and the carbonate mineral content should be measured.

From the site data, both spatial and temporal trends should be identified. These trends can be used to aid in the identification of important reactions in the system and to quantify the current mobility of contaminants. Spatial trends include both the horizontal and vertical distribution of contaminants downgradient of the source area. Temporal trends consist of changes in contaminant level observed by periodic monitoring at a particular location. Information on the timing and level of release of contaminants to the environment is very useful in evaluating spatial and temporal trends. For example, if a relatively nonreactive contaminant such as nitrate was added to the system for a short period, its movement would not be expected to be significantly retarded by adsorption/desorption reactions. The total mass of nitrate in the aquifer may be contained in a plume with a size limited by the duration of contamination with some spreading due to natural dispersion. If a comparison between the original mass of nitrate contamination added to the aquifer and the current mass shows an appreciable loss of nitrate, then denitrification may be occurring to lower solution concentrations. In addition, denitrification may produce ammonium, NH_4^+, which does participate in adsorption/desorption reactions. Ammonium adsorption would significantly retard and spread out the plume of nitrogen contamination due to both nitrate and ammonium forms of nitrogen. Accurate information on the amount of contaminant added to the system and its current distribution are required to identify aquifer processes such as these.

With a reasonable understanding of the geochemical system to be restored, a reaction model for the site can be developed. This model takes into account all of the important water/rock interactions from the standpoint of contaminant sources and mobility. Figure 8-3 shows a simplified conceptual model of a site impacted by acid leachate. Data from the natural system upgradient of the source area can be used to characterize the reactive mineral phases (carbonates, clays) and the groundwater properties (Ca–HCO_3^- type, neutral pH, oxidizing, low total dissolved solids level and metals concentrations). The impact of the leachate on this system can be seen in a transect downgradient of the source area. Beneath the tailings pile and for some distance downgradient, the acid in the leachate has overcome the acid-neutralizing capacity of the system and the pH has dropped producing an acidic zone in the aquifer. Calcite has dissolved releasing calcium, which has combined with the

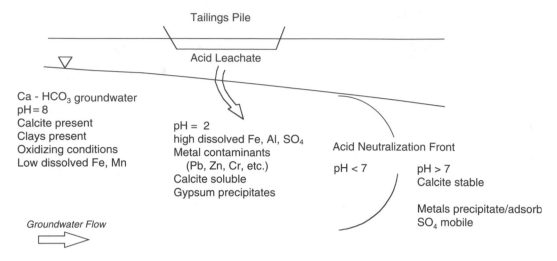

Figure 8-3   Data and conceptual geochemical model of acid-contaminated site.

sulfate from the sulfuric acid source to precipitate gypsum. The high solubility of gypsum, especially in the presence of high levels of dissolved complexing agents such as Al and Mg, maintains elevated sulfate levels (>10,000 mg/L) in solution. Metals such as Fe, Al, Pb, and Zn may also be present in the acidic zone of the plume because of the high solubility of their minerals under these conditions. At the acid neutralization zone, the carbonate minerals present in the aquifer are sufficient to raise the pH back to more neutral values. Under these conditions, the minerals containing metals are much less soluble and the metals are consequently less mobile and may not pass this zone at a concentration level of concern. However, sulfate may continue to exist in solution at a level greater than 2000 mg/L because the solubility of gypsum is not significantly affected by pH. In addition, the precipitation of gypsum in the aquifer will provide a long-term source of sulfate to the groundwater that will likely extend the effort required to achieve remediation in this case. Accurate data on the site is necessary to understand the distribution of contaminants and other compounds that impact their mobility and to take into account geochemical processes that may extend the restoration program.

Working Toward Stable Chemical Equilibria

The introduction of contamination into an aquifer disturbs the geochemical equilibrium of the natural system, and as discussed in Chapter 7 the system reacts in the direction to establish a new equilibrium that accommodates the contaminant. Depending on the degree of disequilibrium produced by contamination and the rate at which the system reacts, equilibrium may or may not be achieved in the system at the time that restoration commences. When an active restoration method is applied to a site, it also perturbs the geochemical system. If this perturbation removes or immobilizes the contaminant and restores the original geochemical equilibrium condition, then long-term stability of the system is likely. For example, in the case of *in situ* leach uranium mining, if the restoration method consists of injecting a reductant into the system to restore reducing conditions to the roll front and immobilize uranium as uraninite, then it is likely that this will be a stable condition at the completion of restoration activities because the natural reducing condition of the roll front has been reestablished. At the cessation of injecting the reductant, uraninite will not dissolve and release uranium because the natural system is reducing due to a lack of oxygen at depth in the aquifer.

If the composition of the groundwater upgradient of the restoration zone is not considered in restoration design, then attempting to immobilize a contaminant may be only temporarily successful. For example, the concentration of the contaminant barium can be limited in groundwater by the mineral barite ($BaSO_4$). The limit on barium concentration imposed by barite equilibrium is a function of the sulfate concentration in the groundwater. In a dilute groundwater if the sulfate concentration is greater than a few milligrams per liter, then at barite equilibrium the dissolved concentration of barium will be less than one milligram per liter (see Section 3.2). Thus it is possible to depress the dissolved concentration of barium by adding sulfate to water in a zone of barium contamination. However, if the natural groundwater has low sulfate concentration, restoration will be only temporary. When the natural groundwater with low sulfate concentration reenters the restored zone, the solubility of barite will increase and the dissolved concentration of barium may exceed the cleanup level. In this case the restoration method neglected the long-term consequences of the applied technique as the natural system reestablished itself. In fact, the precipitation of barite in this aquifer would increase the difficulty of restoration because the presence of barite represents a secondary source of barium in the aquifer that must also be removed along with the dissolved barium.

Figure 8-4 depicts a situation in which a pesticide-producing facility has contaminated an aquifer with arsenic. Localized reducing conditions caused by the contaminant solution has also dissolved iron oxyhydroxide minerals in the aquifer and elevated the dissolved iron concentration. In this case the natural condition of the aquifer is oxidizing, and reestablishing oxidizing conditions in the zone of contamination may immobilize the contaminants. If oxygen is injected into the aquifer to raise the redox potential, the iron will precipitate as iron oxyhydroxide minerals, which are relatively insoluble under the natural oxidizing conditions. These minerals also have a strong affinity for arsenic and will scavenge arsenic from solution and immobilize it on the mineral surfaces. If the affinity of the mineral for arsenic is strong enough, it may lower the dissolved arsenic concentration to below a level of concern. Reoxidizing the zone of contamination may then restore the aquifer, and because the natural system is oxidizing, the iron and arsenic will remain at low concentrations in solution when the natural groundwater enters the previously contaminated zone at the termination of oxygen injection. For this approach to be effective, the amount of arsenic contamination in solution cannot exceed the adsorption capacity of the precipitated iron oxyhydroxides. Calculations must be made to estimate the amount of iron oxyhydroxides that will be available to scavenge the arsenic.

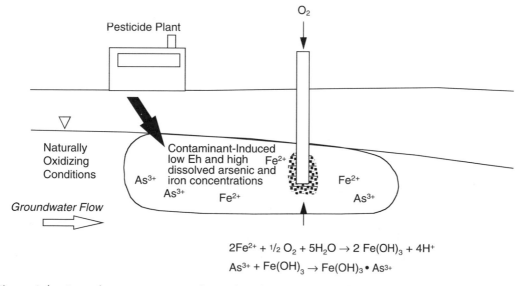

$$2Fe^{2+} + \tfrac{1}{2} O_2 + 5H_2O \rightarrow 2\,Fe(OH)_3 + 4H^+$$
$$As^{3+} + Fe(OH)_3 \rightarrow Fe(OH)_3 \bullet As^{3+}$$

Figure 8-4   Groundwater arsenic remediation by adsorption onto $Fe(OH)_3$.

In evaluating the stable equilibrium condition of the aquifer, it is necessary to know the pH and Eh of the system, as well as the concentrations of the important dissolved species. Rates of reactions may need to be considered for stable minerals that are slow to equilibrate under aquifer conditions. For example, the formation of $Cr(OH)_3$, which may limit chromium concentration in groundwater, is relatively slow at pH values greater than about 6.[8] Either sufficient time must be allowed for reactions to come to equilibrium or catalysts may be necessary to facilitate the reaction. In the case of organic contaminants, oxygen or nutrients may need to be added to foster bacterial growth and consumption of the contaminants.

Identifying and Avoiding Unwanted Side Effects

Attempting to remove a contaminant from an aquifer may mobilize other contaminants that were not initially at a level of concern. This occurrence can add significantly to the scope of the restoration effort. Figure 8-5 shows the situation where organic contaminants from a leaking underground storage tank have entered an aquifer that contains pyrite (FeS_2). The aquifer is naturally reducing with respect to iron and sulfur. A commonly applied, and successful, remediation technique for hydrocarbon compounds is to create oxidizing conditions in the zone of organic contamination so that aerobic bacteria can consume the contaminants. At the site shown in Figure 8-5 the introduction of oxygen also creates a localized environment where pyrite is unstable. Pyrite will oxidize, producing sulfuric acid. If there is not sufficient neutralizing capacity in the aquifer, the pH of the system will drop, increasing the mobility of many metals. The original organic contamination problem has now turned into an organics, metals, and pH problem, and it may be more difficult to cleanup the inorganic contamination than the original organic contamination.

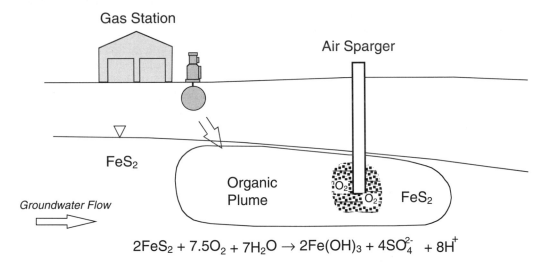

$$2FeS_2 + 7.5O_2 + 7H_2O \rightarrow 2Fe(OH)_3 + 4SO_4^{2-} + 8H^+$$

Figure 8-5   Aerobic bioremediation-induced acidic conditions.

Potential detrimental side effects to a remediation effort include (1) clogging of pores and reduced transmissivity due to mineral precipitation and bacterial growth, (2) generation of unwanted gases such as CO_2 and H_2S, and (3) mobilization of contaminants initially not at a level of concern but which become elevated in groundwater because of desorption or increased solubility of a mineral containing the contaminant. Common minerals that might precipitate include the carbonates and the oxides and hydroxides of iron and manganese. Bacterial growth may accompany the oxidation of organic contaminants and the oxidation/reduction of iron, nitrogen, and sulfur species. To be able to predict the likelihood of these occurrences, the geochemical system must be well characterized and all possible detrimental reactions must be evaluated. This evaluation generally consists of developing a conceptual geochemical model of the contaminated system and then simulating the impact of restoration on this system using one of the modeling codes described in Chapter 5.

8.3. APPLIED RESTORATION TECHNIQUES

If the natural system is not capable of lowering contaminant concentrations to below a level of concern or if the rate of natural processes are not fast enough, then active restoration measures may be required. Commonly applied restoration techniques can be grouped into two categories: pump-and-treat and *in situ* treatment methods. The geochemistry of the contaminated system may inhibit or enhance these methods.

Pump-and-Treat Methods

The concept of the pump-and-treat method is straightforward. If contamination has entered an aquifer, producing high levels of a substance in the groundwater, then removing the contaminated groundwater by pumping should restore the aquifer. Surface treatment of the contaminated water may be required before discharge. Unfortunately, application of this simple technique for aquifer remediation has rarely been successful. The reasons for failure have been attributed to both physical and geochemical processes active in the aquifer. The physical process of dispersion spreads out the contamination and lowers extraction efficiency. The diffusion of contaminants from less permeable flow zones and deadend pores will also extend the amount of time required for flushing the aquifer of dissolved contaminants. The primary geochemical processes that inhibit the effectiveness of contaminant removal are adsorption/desorption and mineral dissolution reactions.

The potential impact of these physical and geochemical processes on the pump-and-treat process is depicted in Figure 8-6. These processes produce a phenomenon termed *tailing* in which the concentration of the contaminant in the extracted groundwater, which initially may drop quickly during the early stages of pumping, reaches a concentration plateau during which time pumping does not effectively lower the contaminant level.[9]

The most common geochemical process that might produce this type of response is the dissolution of a mineral containing the contaminant. If the contaminant is contained in

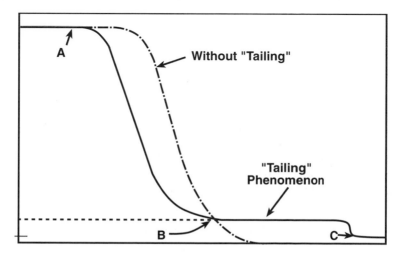

Figure 8-6   Concentration vs. time for an extraction well with continuous pumping. (From Palmer, C.D. and Fish, W., *Chemical Enhancements to Pump-and-Treat Remediation*, U.S. Environmental Protection Agency, 1992.)

a reactive mineral, then the removal of the contaminant from the system will be limited by the solubility of that mineral. For example, consider a system contaminated with Pb in which the lead carbonate mineral cerrusite ($PbCO_3$) has precipitated. If, during the contamination phase, the flux of Pb to the aquifer exceeds the flux of carbonate, then precipitation of cerrusite will limit the carbonate concentration of the groundwater while Pb concentrations will be elevated. Cerrusite forms as a solid-phase source of Pb in the aquifer, but its formation does not limit the Pb concentration in the groundwater because Pb is present in excess of the carbonate component of the mineral. During the initial stages of remediation by pump-and-treat it is expected that high concentrations of Pb will be observed in the extracted water (region A on Figure 8-6), but once this highly-contaminated water has been removed, the Pb concentrations will drop relatively quickly to point B on Figure 8-6. At this point cerrusite becomes the principal contributor of Pb to the groundwater, and, depending on its solubility under the pH and carbonate concentrations of the groundwater being drawn into the contaminated zone, the Pb concentrations may remain above a level of concern. If this is the case, all of the cerrusite will have to be dissolved and removed from the aquifer before the Pb concentrations in the extracted water can decrease to a lower level (point C on Figure 8-6). To estimate the amount of time and volume of water that must be pumped through the aquifer to achieve remediation when a mineral is elevating the contaminant concentration level, the amount of the mineral present in the aquifer must be known, as well as the solubility of the mineral under the restoration conditions.

Desorption is the other geochemical process that may significantly increase the amount of pumping required to remove a contaminant from an aquifer. Figure 8-7 depicts an initial contaminated situation in which Cr(VI) is present in solution and adsorbed on the surface of the aquifer solids. The site-specific Kd for chromium is 2 L/kg. The initial solution concentration is 1 mg/L and the solid-phase, adsorbed concentration is 2 mg/kg. Under these conditions the total mass of Cr in a liter of water is 1 mg and the total mass of Cr adsorbed to the solids in contact with a liter of water is 16 mg (assuming a bulk density of 2 g/cc and a porosity of 0.25). If all of the contaminated water was replaced with Cr-free water during the first flush, then Cr would desorb off of the solids to re-equilibrate with the flush water. At a Kd value of 2 L/kg, the resulting solution and solid phase Cr concentrations are 0.94 mg/L and 1.88 mg/kg, respectively (Figure 8-7). The Cr concentration in solution did not decrease very much because of the large reservoir of Cr on the surface of the aquifer solids compared with the mass of Cr in solution. Figure 8-8 shows the results of several flush cycles on the concentrations of Cr in solution and on the solid. It shows that five flushing cycles reduce the groundwater concentration by only 0.26 mg/L.

The impact of desorption on the efficiency of pump-and-treat is directly related to the affinity of the surfaces for the contaminant. The stronger the affinity of the solid for the contaminant (higher Kd values), the greater the number of flush cycles required to lower contaminant levels in an aquifer. Figure 8-9 shows the effect on chromium removal under situations where its Kd is 0.5 and 2 L/Kg. For the Kd of 0.5 L/kg, five flush cycles lower the solution concentration from 2 to 0.66 mg/L, whereas for a Kd of 2 L/kg the groundwater concentration is only lowered from 1 to 0.74 mg/L. The general relationship between number of flush cycles (n) and Kd to lower groundwater concentrations from an initial value (C_i) to a desired cleanup level (C) is as follows:

$$C/C_i = \left(1 - 1/R\right)^n \qquad (8-1)$$

where R is the retardation factor $(1 + (\rho_b/\eta)Kd)$

Table 8-1 provides calculated values of the number of flush cycles for various ratios of C/Ci and R. As can be seen, even a relatively small affinity of the solid for the contaminant (low R value) can substantially increase the amount of pump-and-treat required. Also,

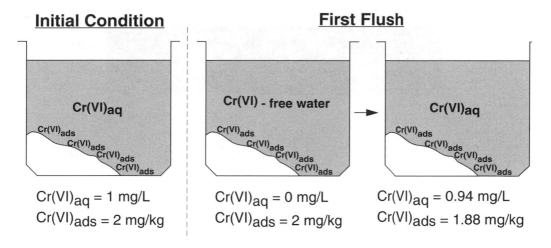

Figure 8-7   Chromium desorption (Kd = 2 L/kg).

Number of Flushes	Cr(VI)aq (mg/L)	Cr(VI)ads (mg/kg)
0	1	2
1	0.94	1.88
2	0.88	1.77
3	0.83	1.67
4	0.79	1.57
5	0.74	1.48

Figure 8-8   Simulation of chromium flushing from an aquifer.

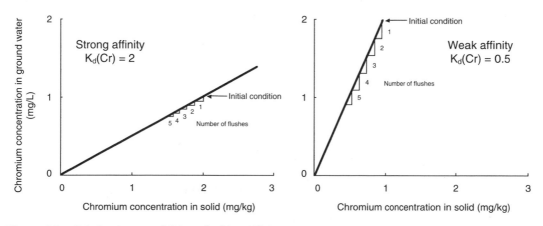

Figure 8-9   Relative impact of Kd on flushing efficiency.

increasing the amount of contaminant reduction required (decreasing C/Ci) has a smaller effect on the number of pore volumes needed to flush through the system to achieve a cleanup value than increasing the retardation factor. These calculations assume a linear isotherm in which the affinity of the solid for the contaminant is constant at all concentrations. If desorption follows a nonlinear isotherm such as a Freundlich or Langmuir isotherm

TABLE 8-1.

**RELATIONSHIP OF NUMBER OF FLUSH
CYCLES (n) TO RETARDATION FACTOR (R)
AND REQUIRED DEGREE OF CONTAMINANT
REDUCTION (C/C$_i$)**

C/C$_i$	R	n
0.5	5	3
0.1	5	10
0.01	5	21
0.1	10	22
0.1	20	45
0.1	30	68

(Section 3.1), then at lower solution concentrations the affinity may increase, thereby requiring an additional amount of pumping. These calculations also assume that desorption equilibrium is achieved as fresh water enters the zone of contamination. Additional flushing will also be required if this is not the case.

The overall impact of mineral equilibrium and desorption on the efficiency of pump-and-treat remediation is shown schematically in Figure 8-10. This figure relates the dissolved concentration of a contaminant to its concentration in/on the solid phase. At solution concentrations below saturation with a mineral phase, adsorption/desorption reactions influence solution concentration and provide a source and sink for the contaminant. As solution concentration increases, it may be limited by the formation of a solid containing the contaminant. This is shown on the figure as the mineral equilibrium zone where the concentration line is horizontal. If the groundwater cleanup level is below the mineral equilibrium level and a reactive mineral containing the contaminant is present in the aquifer, then all of the mineral must dissolve before the solution concentration will decrease below the mineral equilibrium line. From this point on, desorption will be a source of the contaminant to the water flushed through the aquifer. The efficiency of flushing during the desorption phase of restoration is related to the slope and shape of the desorption line. If the line is steep, flushing will be more efficient than if the line is flat. If the adsorption isotherm is not straight, but curves upward as in the case of the Freundlich and Langmuir isotherms, then desorption efficiency will decrease as solution concentrations are lowered.

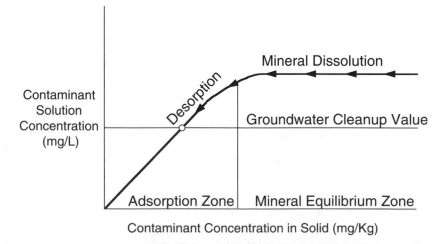

Figure 8-10 Mineral dissolution and desorption to achieve groundwater cleanup.

In Situ **Treatment Methods**

Treatment of aquifer contamination *in situ* is an attractive alternative to pump-and-treat because it minimizes the surface handling of contaminated media and reduces the volume of waste, either solution or solid, requiring disposal. *In situ* methods may involve the addition of treatment chemicals to the zone of contamination or the construction of a geochemical barrier to intercept contaminated water as it flows through the aquifer. In the case of the injection of a treatment chemical, not only must the reactant immobilize or decompose the contaminant to a concentration below a level of concern, but it must generate a stable geochemical environment in which the contaminant will not be remobilized at some later time. Injection of a reactant chemical also must not produce deleterious side effects (see Section 8.2). Reactant injection may be as simple as adding oxygen to a plume of organic contamination to enhance bioremediation or may be more sophisticated, such as adding dissolved iron that precipitates in the aquifer and scavenges cations and anions or introducing a reductant into the aquifer to immobilize metals.

In the case of the uranium *in situ* leach mine site discussed in the context of natural restoration (Section 8.1), an active treatment method that has been tested in the laboratory consists of adding sodium sulfide (Na_2S) to water injected into the leach zone at the termination of mining (Figure 8-11). The intent is to reproduce reducing conditions in the aquifer and immobilize the uranium and the other potential contaminants (As, Se, and Mo) that may also be mobilized by the mining operation. As shown in Figure 8-11, the Na_2S dissolves in the water to form the strong reductant bisulfide (HS^-). Bisulfide consumes any available molecular oxygen (O_2), thereby lowering the redox potential of the system. Additional bisulfide reacts with dissolved U(VI) species reducing the uranium to U(IV) and precipitating it as uraninite (UO_2), which has a low solubility under reducing conditions. Bisulfide will also react with Fe(III) oxyhydroxide minerals converting them to mackinawite (FeS) and ultimately pyrite (FeS_2). By adding bisulfide to the system, the uranium has been immobilized by a process similar to that in which it was originally concentrated at the roll-front deposit. Also, the formation of FeS and FeS_2 provides a long-term reducing capacity to this zone in the aquifer, which will keep conditions reducing and uraninite stable. Laboratory experiments[10] have shown the potential validity of this approach to *in situ* treatment at these sites.

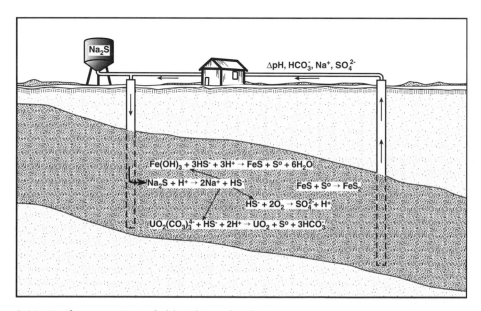

Figure 8-11 Aquifer restoration aided by chemical reduction.

The use of reactive material placed along the flow path of the contaminants has been suggested as part of the funnel-and-gate remediation system.[11,15] A plan view of such a system is shown in Figure 8-12. In this system an impermeable barrier funnels the contaminated water to the gate, which is permeable and contains material that will react with the contaminant, either immobilizing it (metals) or converting it to a less hazardous form (in the case of NO_3^- and organic contaminants). Potential reactive media and the contaminant impacted include (1) reduced iron (Cr, trichloroethene, dichloroethene); (2) crushed limestone (metals, neutralization of low-pH water); (3) sawdust (NO_3^-); (4) NO_3^-/PO_4^{3-} with peat (BTEX); and (5) oxygen-release compound (BTEX). Many factors need to be considered in the design of such a system to adequately treat the contaminant in what may be a short contact time with the reactant and to minimize plugging of the gate. Provisions must be made for replacing the gate material if the reactant is totally consumed or plugging has lowered its efficiency. However, the concept of treating contaminants *in situ* with an engineered system that not only restricts plume expansion but consumes or concentrates the contaminants in a small volume is worth evaluating at sites amenable to this approach.

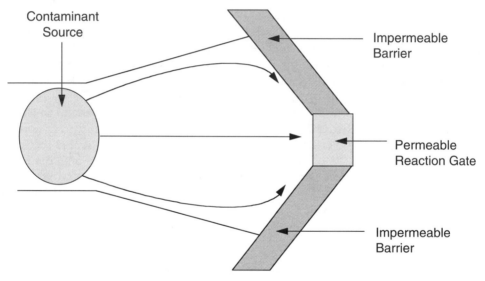

Figure 8-12 Funnel-and-gate remediation system (plan view).

8.4. GEOCHEMICAL ENHANCEMENTS TO PUMP-AND-TREAT RESTORATION

In a sense, *in situ* treatment methods are geochemical enhancements to natural restoration at a site. In a similar manner, the pump-and-treat restoration method can benefit from geochemical manipulation to increase the efficiency of contaminant removal. The two major types of geochemical enhancements produce conditions that either (1) increase the solubility of a reactive mineral containing a contaminant or (2) desorb contaminants by adding dissolved species that compete for the adsorption sites or dissolve the adsorbent, releasing the contaminants to the groundwater. The goal of the pump-and-treat technique is to remove contaminants from the aquifer; therefore geochemical methods that immobilize the contaminant are not considered enhancements to pump-and-treat but rather in situ treatment methods.

Mineral Solubility Enhancement

Increasing the solubility of a mineral containing a contaminant during pump-and-treat restoration will temporarily increase contaminant solution concentrations, thereby enhancing removal of the contaminant from the solid phase. The solubility of a mineral may be affected by many solution parameters including pH, Eh, type and concentration of complexing species, and ionic strength of the groundwater.

Because the components of some mineral types form strong complexes with hydrogen (e.g., carbonate and phosphate) and others form strong complexes with hydroxyl (e.g., Fe and Al), the solubility of minerals containing these components will be strongly dependent on groundwater pH. For example, the solubility of cerrusite ($PbCO_3$) in dilute water increases by a factor of 2 as the pH is lowered from 8 to 7 and increases by a factor of 4.5 as the pH is lowered from 7 to 6. A similar groundwater pH change would have very little effect on the solubility of a sulfate mineral such as barite ($BaSO_4$) because barium and sulfate do not form strong complexes with H^+ or OH^- in the pH range of normal groundwater.

The major potential challenges with pH manipulation to enhance contaminant mobility are the buffering capacity of the system and the possibility of unwanted mobilization of new contaminants. The generally strong pH buffering capacity of many geochemical systems will require large amounts of acid or base to be added to the system to significantly modify its pH. If the pH is changed, it may not only mobilize the contaminant of interest but also other contaminants that were not initially above a concentration level of concern in the aquifer. Another potential complicating factor to consider is the precipitation of a second reactive mineral containing the contaminant. As the solution concentration of the contaminant increases because the solubility of the original mineral increases, the dissolved contaminant concentration may exceed the solubility of another mineral that contains the contaminant, and the precipitation of that mineral may then become the solid limiting the removal of the contaminant. For example, if lead is mobilized by adding sulfuric acid (H_2SO_4) to the aquifer thereby increasing the solubility of cerrusite ($PbCO_3$), saturation may be reached for the lead sulfate mineral anglesite ($PbSO_4$). Anglesite may then precipitate limiting dissolved Pb concentration in solution. In evaluating the effect of pH (or other parameter) manipulations on contaminant mobility, all possible reactions need to be considered.

The solubility of minerals containing redox-sensitive contaminants such as As, Se, Cr, Cu, and Hg will be a function of the redox state of the aquifer. For example, the solubility of the As(V) mineral scorodite ($FeAsO_4 \cdot 2H_2O$), which contains two redox-sensitive components Fe and As, increases by a factor of 4.6 as the redox potential is lowered from +300 to +200 mV and increases by 12x as the Eh decreases from +200 to +100 mV at a pH of 7. The major influence on the solubility of scorodite is the shift of dominant iron redox species from Fe(III), the component in scorodite, to Fe(II). This decreases the concentration of Fe(III) in solution and requires more scorodite to dissolve to reach equilibrium. As with pH manipulation, the capacity and reactivity of the system to buffer the Eh at its natural value must be considered in designing an Eh manipulation process. In addition, because redox reactions can be sluggish, the rates of oxidation and reduction must be considered. Eh manipulation can be produced in an aquifer by adding oxidizing agents such as oxygen gas, atmospheric air, or hydrogen peroxide to raise the redox potential or reducing agents such as reduced iron [Fe(0), Fe(II) or sulfur species [S(0), S(-II)] to lower the redox potential.

The solubility of any mineral can be increased by adding a complexing agent(s) to the solution thereby lowering the effective concentration of the dissolved components of the mineral. The result is that more of the mineral must dissolve to reach equilibrium dissolved concentrations. Many organic compounds such as EDTA[12] and citric acid[16] are known to be strong complexers for metals; however, the injection of an organic compound into an aquifer for remediation of metals may create its own cleanup problems in the future. Naturally occurring, strong inorganic complexants such as the divalent major cations (Ca^{2+} and Mg^{2+}) and anions (SO_4^{2-} and CO_3^{2-}) in groundwater can be considered for their impact on mineral solubility. For example, the solubility of barite ($BaSO_4$) can be increased by adding Mg^{2+} to the water in contact with the mineral. The Mg^{2+} complexes with some of the SO_4^{2-} released from the dissolving mineral and forms the neutral solution species

$MgSO_4^0$. The solubility of barite can be increased by a factor of 1.3 by adding 0.001 moles/L Mg^{2+} (24 mg/L) to the solution, and it can be increased by a factor of 2.3 by adding 0.01 moles/L Mg^{2+} (240 mg/L). The ability to remove the complexant from the system or reduce it to an acceptable concentration at the cessation of restoration is an important design consideration, which is one reason why the naturally occurring major ions are attractive alternatives.

As discussed in Chapter 3 on mineral equilibrium, ion shielding can be an important solution process that affects mineral solubility. Like the complexation reactions, ion shielding lowers the effective concentration of dissolved species and increases mineral solubility. The ionic strength of a solution is a measure of its ion shielding ability; therefore increasing the ionic strength of the solution pumped through the zone of contamination will increase the solubility of all reactive minerals. The only exception occurs if the components of the solution used to increase ionic strength are also present in the minerals. In this case the solution would depress mineral solubility. For this reason relatively nonreactive ions such as sodium and chloride would be the preferred choice for increasing ionic strength without inhibiting mineral dissolution. In the case of barite it was shown in Table 3-10 that mineral solubility can be increased by factors of 1.5 to 2 for each tenfold increase in ionic strength produced by the addition of Na^+ and Cl^- to the groundwater.

Contaminant Desorption Enhancement

The presence of an adsorbed contaminant on the surface of aquifer solids provides a continuing source of contamination to the groundwater until enough of the contaminant has been desorbed that the solution concentration in equilibrium with the surface concentration is below the cleanup level. Depending on the amount of contaminant that must be removed from the solid and the affinity of the solid for the contaminant, this may require an excessive amount (hundreds of pore volumes) of pumping to achieve groundwater cleanup. Desorption can be enhanced by adding a dissolved constituent that competes with the contaminant for the surface sites, or a parameter of the system such as pH or Eh can be changed to decrease the affinity or capacity of the surface for the contaminant.

It has been well documented in the soil chemistry literature that clay minerals with their predominantly anionic surface charge under neutral and higher pH values are strong cation adsorbers with a preference for divalent cations (Ca^{2+}, Mg^{2+}, Pb^{2+}) over monovalent cations (Na^+, K^+, Cu^+).[13] The affinity and capacity of other adsorbents in an aquifer such as the metal oxyhydroxides is highly pH dependent. For instance, ferrihydrite favors the adsorption of anions below a pH of about 8; however, it does specifically adsorb cations at lower pH values. By using either site-specific selectivity coefficients for the adsorption of cations onto clays or the compiled adsorption constants for cations and anions onto ferrihydrite, decisions can be made as to the best species to introduce into an aquifer to enhance desorption of the contaminant.

Desorption can be enhanced not only by choosing a species with a stronger affinity for the surface but by introducing it at higher dissolved concentrations than the contaminant. If the only viable species has a lower affinity than the contaminant, then enhancement must be optimized by increasing the concentration of competing species accordingly. If this requires too high a dissolved concentration level, then this method of enhancing desorption would not be practical. For example, arsenic strongly adsorbs onto ferrihydrite as an anion. Sulfate may be considered as an anion to compete for the sorption sites containing arsenic; however, arsenic anions are generally much more strongly held by ferric hydroxide than sulfate. Therefore very high concentrations of sulfate in solution would be required to remove the arsenic.

When desorption cannot be practically enhanced by adding a competing ion to the solution, more direct methods of modifying the adsorbent can be considered. The sign of the surface charge on the adsorbents is a function of pH.[14] At a pH value below the point-of-zero charge (PZC) the surface charge is predominantly positive and anions are attracted and adsorbed to the surface, while at a pH above this point the surface is negative and the adsorbent preferentially attracts positively charge species. Table 3-3 shows the PZCs for commonly important adsorbents in aquifer systems. The fact that the clay minerals have PZCs less than 5, which is below the typical groundwater pH value, is the reason clays have high cation exchange capacity and low anion exchange capacity. By changing the pH of groundwater in the contaminated zone it is theoretically possible to enhance desorption. This may not be practical for cations adsorbed onto clays because the pH would have to be lowered to such a low value that a large amount of acid would be required and other unwanted reactions would probably also occur. In the case of ferrihydrite where the PZC of about 8 is in the range of natural groundwater, pH manipulation may be a viable enhancement technique if the system is not well buffered and other reactions do not inhibit the desired effect.

A final method of enhancing desorption would be to dissolve and remove the adsorbent from the aquifer. Although this may not be practical for the clay minerals, it may be for the Fe and Mn oxyhydroxide minerals, which are strong adsorbers for trace metals. The solubilities of these solids are strongly pH and Eh dependent. If the aquifer conditions can be changed enough to dissolved these solids, they will release all adsorbed contaminants allowing them to be flushed from the aquifer by the pump-and-treat operation. The initial pH and Eh of the aquifer and its pH buffering and Eh poising capacities will determine the amount of reactant required to solubilize the metal oxide and hydroxide solids.

8.5. SIMULATING THE GEOCHEMISTRY OF RESTORATION

Geochemical reaction modeling provides a tool for evaluating the extent of natural restoration that a system is capable of attaining and for testing the effectiveness of *in situ* techniques and geochemical enhancements to the pump-and-treat method. The requirements for developing a restoration simulation model are sufficient data to generate a valid conceptual model of the contaminated site and a method of calculating the important geochemical processes that might occur during proposed restoration alternatives. The MINTEQ and PHREEQE computer codes described in Chapter 5 are suitable for this purpose when reaction kinetics are not a limiting factor. To use these thermodynamic codes to evaluate remediation, the reactions must achieve equilibrium conditions or the results will be purely qualitative. This is not generally a major concern in remediation modeling because the reactions chosen to remediate a site should be relatively fast to have an effect in a typical restoration time frame measured in months or, at most, years.

The calculations of the impacts of pH, Eh, complexants, and ionic strength on mineral solubility described in Section 8.4 were performed with the MINTEQA2 computer code. The data requirements to perform these calculations were simply the complete solution chemistry of the groundwater and the equilibrium constant of the mineral limiting solution concentration of the contaminant. In addition to calculating mineral solubility under various aquifer conditions, the model of the system can also be used to estimate the amount of reagent that must be added to the system to achieve the desired change in pH, Eh, etc.

Figure 8-13 shows the results of modeling changes in pH and Eh to enhance the desorption of arsenic from ferrihydrite. The MINTEQA2 code was also used for this simulation. The contaminated aquifer has a pH of 5.25 and an Eh of +300 mV. Arsenic is adsorbed onto the solid and has an equilibrium solution concentration of 5.5 mg/L. In the first case the redox potential is held constant, and the pH is increased in increments of one unit to a final value of 8.25. As the pH of the solution increases, the surface of the ferrihydrite becomes less positively charged, thereby decreasing its capacity to adsorb the arsenic anion. As shown on the table in Figure 8-13, about seven times as much arsenic is expected in solution at a pH of 8.25 than at a pH of 5.25. Increasing the pH has a large effect on mobilizing and removing arsenic for this system. In the second simulation the pH

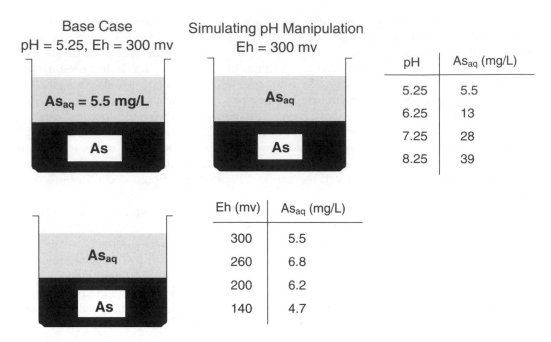

Figure 8-13 Simulating remediation of arsenic contamination.

was held constant, and the Eh was decreased from +300 to +140 mV. Changing the redox potential over this range did not have an appreciable effect on enhancing the mobility of arsenic, probably because Eh manipulation does not significantly change the surface properties of the ferrihydrite and the Eh is not low enough to change the dominant arsenic redox species, As(V).

The variety of simulations that can be quickly tested with these computer models is only limited by the imagination of the investigator and the capabilities of the computer code. Before implementing a restoration technique that appears promising in the simulations, it is necessary to evaluate the method in the lab where actual site materials can be tested and conditions can be controlled. If the results agree with the model calculations, then a pilot field study is warranted to provide engineering data to scale up to full-scale remediation. Unforeseen occurrences along the developmental pathway may warrant a return to the modeling stage and modification to the conceptual model. The iterative process improves understanding of the system to be remediated and increases the likelihood of eventual success.

REFERENCES

1. Perel'man, A., Geochemical barriers: theory and practical applications, *Appl. Geochem.*, 1, 669–680, 1986.
2. Moore, J.N., Ficklin, W.H., and Johns, C., Partitioning of Arsenic and Metals in Reducing Sulfidic Sediments, *Environ. Sci. Technol.*, 22, 432–437, 1988.
3. Applin, K.R. and Zhao, N., The kinetics of Fe(II) oxidation and well screen encrustation, *Ground Water*, 27, 168–174, 1989.
4. Baedecker, M.J. and Back, W., Hydrogeological processes and chemical reactions at a landfill, 17, 429–437, 1979.
5. Davis, A., Kempton, J.H., Nicholson, A., and Yare, B., Groundwater transport of arsenic and chromium at a historical tannery, Woburn, Massachusetts, U.S.A., *Appl. Geochem.*, 9, 569–582, 1994.

6. Galloway, W.E. and Kaiser, W.R., *Catahoula Formation of the Texas Coastal Plain: Origin, Geochemical Evolution, and Characteristics of Uranium Deposits*, Bureau of Economic Geology, The University of Texas at Austin, 1980.

7. Deutsch, W.J., Martin, W.J., Eary, L.E., and Serne, R.J., *Methods of Minimizing Ground-Water Contamination from In Situ Leach Uranium Mining*, Pacific Northwest Laboratory, 1985.

8. Henderson, T., Geochemical reduction of hexavalent chromium in the Trinity Sand Aquifer, *Ground Water*, 32, 477–486, 1994.

9. Palmer, C.D. and Fish, W., *Chemical Enhancements to Pump-and-Treat Remediation*, U.S. Environmental Protection Agency, 1992.

10. Deutsch, W.J., Eary, L.E., Martin, W.J., and McLaurine, S.B., *The Use of Sodium Sulfide to Restore Aquifers Subjected to In Situ leaching of Uranium Ore Deposits*, National Ground Water Association, Reno, Nevada, 1985.

11. Starr, R.C. and Cherry, J.A., In situ remediation of contaminated ground water: the funnel-and-gate system, *Ground Water*, 32, 465–476, 1994.

12. Vuceta, J. and Morgan, J.J., Chemical modeling of trace metals in fresh waters: role of complexation and adsorption, *Environ. Sci. Technol.*, 12, 1302–1309, 1978.

13. Brady, N.C., *The Nature and Properties of Soils*, MacMillan Publishing, New York, 1974.

14. Parks, G.A., in *Equilibrium Concepts in Natural Water Systems*, American Chemical Society, Washington, DC, 1967.

15. Gillham, R.W. and D.R. Burris, *Recent Developments in Permeable In Situ Treatment Walls for Remediation of Contaminated Groundwater*, Third International Conference on Groundwater Quality Research, Dallas, TX, June 21–24, 1992.

16. James, B.R. and R.J. Bartlett, Behavior of Chromium in Soils. Fate of Organically Complexed Cr(III) Added to Soil, *J. of Gnv. Qual.*, 12, 119–172, 1983.

9 GEOCHEMISTRY AND THE DESIGN OF SAMPLING PROGRAMS

Data collected at a site are the primary information used to characterize a natural system and evaluate contaminant level, distribution, and mobility as part of a human health or ecological risk assessment. The site data are also one of the most important components of remedial design. Because of the reactivity of the groundwater, solids, and gases collected during sampling, potential geochemical interactions between these phases must be included in the design of sampling programs to ensure that the data accurately represent *in situ* aquifer conditions. This chapter describes the potential impact of geochemical reactivity on data collection and provides general guidelines for developing effective sampling programs. Specific sampling issues that affect particular types of contaminated sites are described in the following chapters that focus on common contaminated environments (e.g., landfills, acid mine drainage sites, and aquifers impacted by organic compounds).

Groundwater and solid-phase data requirements for understanding the geochemistry of a site were discussed in Section 4.1.1 and listed in Tables 4-1 and 4-2. The solution-phase characterization requires data on temperature, pH, Eh, dissolved gases, major cations/anions, trace elements, and organic compounds. Solid phase characterization can be limited to those solids that are reactive under the site environmental conditions. The solids may be reactive because they dissolve or precipitate in the aquifer (e.g., carbonate, sulfate, and sulfide minerals) or they participate in ion exchange or adsorption/desorption reactions (e.g., clay and metal oxide/hydroxide minerals). The impact of sampling on the stability of these aquifer characteristics is described in the following sections.

9.1. TEMPERATURE

Although not a chemical characteristic of the system, many of the geochemical properties (such as mineral and gas solubility) are temperature sensitive; therefore an accurate measure of aquifer temperature is necessary to interpret the chemical data in terms of interactions between the water, solid, and gas phases. As with the other solution parameters that cannot be preserved for later measurement, temperature should be measured in the field, preferably downhole in the well or in a flow-through chamber that minimizes contact with the atmosphere and equilibration with earth surface conditions. Figure 9-1 shows a schematic of a flow-through measurement system that can be attached to a pump discharge line.

9.2. HYDROGEN ION ACTIVITY (PH)

An accurate measurement of groundwater pH is essential because many of the solution processes (aqueous complexation), water/rock interactions (mineral solubility and adsorption properties), gas solubilities, and biochemical reactions are pH sensitive. As discussed in Chapter 1, the pH of groundwater is commonly buffered by the inorganic carbon constituents in solution; however, the pH may still change appreciably when the water is exposed to surface conditions and allowed to equilibrate with atmospheric CO_2 concentrations (Chapter 2).

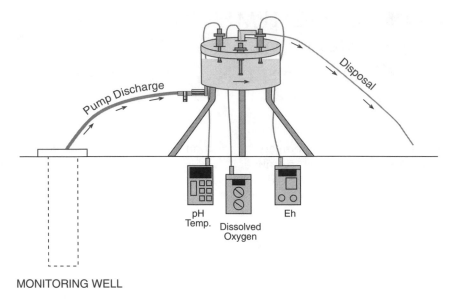

MONITORING WELL

Figure 9-1   Schematic of flow-through measurement system.

Figure 9-2 shows the calculated pH of a groundwater system in equilibrium with calcite at various CO_2 gas partial pressures (P_{CO_2}). At high CO_2 partial pressures, the equilibrium pH is relatively low (6.5 to 7.0) because of the high concentration of carbonic acid produced by the large amount of CO_2 dissolved in the water. At lower CO_2 partial pressures, less carbonic acid is present and the pH in equilibrium with calcite is higher. It is not uncommon for groundwater to be in equilibrium with a high P_{CO_2} because of the oxidation of organic matter in the subsurface. If a sample of groundwater originally in equilibrium with calcite and a P_{CO_2} of 0.1 atmospheres (log P_{CO_2} = –1) is collected and allowed to equilibrate with atmospheric air (log P_{CO_2} = –3.5 atm), the pH of the water would rise from the aquifer

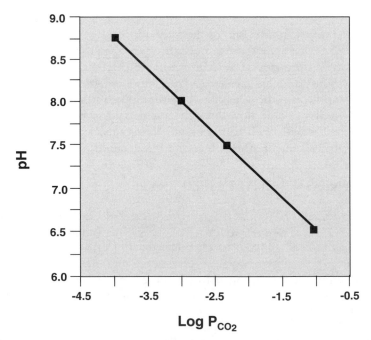

Figure 9-2   Effect of co_2 gas pressure on groundwater pH (CO_2/calcite/H_2O system).

value of 6.5 to about 8.4. This large change in pH would have a major effect on the solubility of minerals. If calcite was initially in equilibrium with the water at a pH of 6.5, it would be oversaturated at a higher pH and would precipitate in the sample bottle. For the calcite/water/CO_2 system, the solubility of calcite decreases by about a factor of 8 as the equilibrium P_{CO_2} decreases from 0.1 to $10^{-3.5}$ atmospheres and the pH increases from 6.5 to 8.4. Metal hydroxide minerals are also generally less soluble at higher pH values and may also precipitate in the sample bottle if they become oversaturated. This is one reason why water samples collected for metals analyses are acidified to a pH of 2. At this low pH, the metals will not precipitate from solution and will be available for measurement by the solution analytical technique.

Some very deep aquifers in which the groundwater has not been in contact with a gas phase containing CO_2 for a long period may be undersaturated with respect to CO_2 gas in atmospheric air. As a result of water/rock interactions, the original dissolved inorganic carbon in the groundwater has been removed by the precipitation of carbonate minerals. The result is that the calculated equilibrium P_{CO_2} is less than the atmospheric value of 0.0003 atmospheres. Some of the deeper aquifers on the Columbia Plateau of eastern Washington State have calculated P_{CO_2} values of 10^{-6} atm. If samples of this water are left exposed to atmospheric air conditions, CO_2 will dissolve into the water to equilibrate with the higher air P_{CO_2}, and the pH of the water will decrease from the actual aquifer value. Although this may not cause minerals to precipitate in the sample container, accurate saturation indices cannot be calculated from an incorrect pH value.

In addition to CO_2 gas exchange, other reactions, such as oxidation-reduction described below, can affect the pH of a water sample. For these reasons a laboratory-measured pH value is usually not acceptable, and the field-measurement method must ensure that equilibration with atmospheric air does not occur.

9.3. REDOX POTENTIAL (EH)

As discussed in Chapter 2, many of the common contaminants in groundwater systems are redox-sensitive (for example, As, Se, Cr, N, organic compounds), and some of the reactions that control their mobility also involve redox-sensitive elements (O, Fe, Mn). Although redox disequilibrium commonly occurs between the various dissolved redox pairs of elements, and the platinum electrode–measured Eh may be representative of only a few of the redox-sensitive elements present in groundwater, the measured redox potential can still be a useful measure of the redox state of the system and the direction that a redox reaction will proceed. For example, as the redox potential of an aquifer decreases due to the reduction of oxidized species by organic matter, it is expected that dissolved oxygen concentrations will decrease, followed by the dissolution of Mn(IV) minerals, the dissolution of Fe(III) minerals, and the reduction of NO_3^- to NO_2^-, N_2, and NH_4^+. At a given Eh, it is possible to predict whether or not the oxidized forms of these redox-reactive elements and their dissolved species or solid phases are stable. For the redox-sensitive elements that are less commonly in equilibrium with the measured Eh (e.g., S), these predictions may not be as reliable because the reactions must be catalyzed by bacteria to reach an equilibrium distribution of the species of the element comprising the redox couple. However, the redox potential still provides a gauge as to which redox species is the most stable in the environment and the direction in which redox reactions may be progressing, albeit slowly, for these elements.

In groundwater systems the Eh is generally less oxidizing than in surface water systems because of the lack of a source of molecular oxygen below the water table and the presence of reducing solids (principally organic matter and sulfide minerals) in the aquifer. However, a sample of groundwater removed from association with these solids and exposed to atmospheric air with its high concentration of oxygen will equilibrate with the air and become more oxidizing, thereby raising the redox potential. In situations where it is anticipated that the aquifer is relatively reducing (Eh < +100 mV) such as beneath and downgradient of bogs/swamps, landfills, and leaky USTs containing organic compounds, it is necessary to take precautions to prevent the water sample from contacting atmospheric air before the redox potential is measured. The water in the well should not be aerated by excessive drawdown during pumping. Also, the pumped water should either be sent through a flow-through chamber containing the measuring electrodes or pumped with minimal agitation into the bottom of a large sampling container with the tip of the electrode at the bottom of the container. The Eh of poorly poised, reduced water can increase several hundred millivolts if these precautions are not taken. An inaccurate Eh prejudices estimates of contaminant mobility and aquifer restoration for redox-sensitive elements because of the Eh sensitivity of mineral solubility and adsorption affinity for these elements.

Surface contamination of a groundwater sample by oxygen in the air may also affect the pH of the sample if it contains high concentrations of a redox-sensitive element that precipitates as the Eh increases. For example, if a reduced groundwater having a pH of 7, an Eh of 0 mV, a dissolved inorganic carbon concentration of 5×10^{-3} moles/L (300 mg/L as CO_3^{2-}), and a total dissolved iron concentration of 30 mg/L is exposed to atmospheric air and allowed to oxidize to an Eh of +400 mV, the iron will precipitate as ferrihydrite. This mineral is relatively soluble at an Eh of 0 mV but insoluble at +400 mV at near neutral pH values. The final total dissolved iron concentration in equilibrium with ferrihydrite at an Eh of +400 mV would be 0.005 mg/L, and the pH would decrease from a value of 7 to 6.34. The pH decreases because the precipitation of ferrihydrite consumes OH$^-$ in the water and releases H$^+$ according to the following reaction:

$$4Fe^{2+} + O_2 + 10H_2O \rightarrow 4Fe(OH)_3 + 8H^+ \qquad (9-1)$$

The presence of inorganic carbon in the water tends to buffer the pH by complexing some of the hydrogen produced by the precipitation of ferrihydrite. This complexation reaction can be written as follows:

$$H^+ + HCO_3^- \rightarrow H_2CO_3 \qquad (9-2)$$

The amount of pH buffering provided by the groundwater is dependent on its dissolved inorganic carbon (DIC) concentration. If the DIC for the water sample given in this example was 3×10^{-3} moles/L (180 mg/L as CO_3^{2-}) instead of 5×10^{-3} moles/L, then the final pH after ferrihydrite precipitation would be 5.95 instead of 6.34. Depending on the buffering capacity of the solution and the presence of reactive, reduced species in solution, the pH of a water sample can be affected by oxidation reactions associated with contact of the sample with atmospheric air. The impact of this reaction can be quite dramatic. A clear water sample from this type of reducing environment can turn reddish over a period of a few minutes to hours after sample collection due to the oxidation and precipitation of the ferrihydrite.

The Eh of a water sample can also be poised just as the pH is buffered; however, Eh poising is usually not as strong as pH buffering in a water sample because of the typically low solution concentrations of redox-reactive elements (Fe, Mn, As, Se) and/or the slow rates of redox reactions. For example, Stumm and Morgan[1] estimate that the Fe^{3+} dissolved concentration should be at least 10^{-5} mol/L (0.56 mg/L) to provide an exchange current at the platinum electrode-solution interface that is great enough to generate a Nernstian Eh (Chapter 2). In some groundwater environments (low pH and/or low Eh) the Fe^{3+} portion of the total dissolved iron concentration is orders of magnitude lower than this value.

9.4. DISSOLVED GASES

Dissolved oxygen is the only gas routinely measured in water samples. The oxygen content of the water is important because of its impact on oxidation-reduction reactions, its use as a qualitative indicator of Eh, and the common requirement to add oxygen to a zone of organic contamination to generate aerobic conditions and facilitate biodegradation. Water in contact with an atmospheric oxygen concentration (P_{O_2} = 0.21 atmospheres) will have a dissolved oxygen concentration that will vary from 14.7 to 7 mg/L over the temperature range of 0 to 35°C. Because of the common disequilibrium between dissolved oxygen concentrations and platinum-electrode Eh measurements, it is not appropriate to calculate a system Eh value from the measured dissolved oxygen concentration.[2,3] Dissolved oxygen is measured in the field by titration techniques or electrode methods. Because of the ubiquitous source of oxygen in air, it is necessary that the oxygen content of groundwater samples be measured with minimal contact between the sample and air in the well bore or on the surface.

Dissolved carbon dioxide [$CO_2(aq)$] is a component of the total dissolved inorganic carbon content of water, which can be calculated from the alkalinity value. The concentrations of the other components of dissolved inorganic carbon are usually of more importance in geochemical calculations than the aqueous concentration of CO_2; therefore it is rarely measured directly in a water sample.

Hydrogen sulfide gas dissolves in water in a manner similar to CO_2 in that a number of dissolved sulfide species (H_2S, HS^-, S^{2-}) are formed. The concentration distribution of these species is pH dependent. Under highly reducing conditions the presence of the sulfide species can have a major impact on metal mobility because of the low solubility of several metal sulfide minerals (e.g., FeS_2, ZnS, PbS). However, instead of measuring the dissolved gas concentration, the total dissolved sulfide concentration is usually measured to calculate the sulfide species and determine the equilibrium status of the sulfide minerals. Measurement of H_2S gas in soil vapor has been shown to be useful in locating subsurface hydrocarbon contamination.[4] Methods for sampling soil gas in the vadose zone are provided by Ballestero et al.[5]

In some situations the dissolved concentrations of reduced gases such as methane (CH_4) and hydrogen (H_2) may be measured in groundwater because exsolution of the gases from the solution may generate explosive concentration levels. In cases where these gases may be present (e.g., landfills and other strongly reducing environments), either the dissolved gas concentrations are measured and the equilibrium partial gas pressures are calculated from Henry's Law constants (Chapter 2) or the solution is allowed to equilibrate with air in a closed system and the resulting gas concentration in the vapor phase is measured directly. In either case care must be taken so that the gases do not escape before the measurements have been completed.

9.5. ALKALINITY

Alkalinity is a measure of a solution's acid-neutralizing capacity. It is generally used to determine the dissolved inorganic carbon (DIC) concentration of a water sample. The DIC is composed of $CO_2(aq)$, H_2CO_3, HCO_3^-, and CO_3^{2-}. Because bicarbonate (HCO_3^-) is usually the dominant anion in shallow groundwaters, an accurate alkalinity measurement is essential to properly characterize the geochemical system. As discussed in Chapter 1, Fritz[6] concluded that the most common sources of charge balance errors for water samples were associated with the alkalinity measurement. Either DIC was lost from the water sample during the alkalinity titration by CO_2 degasing and mineral precipitation, or DIC increased as carbonate mineral particulates in unfiltered samples dissolved into the solution. Furthermore, if other acid-titratable anions (such as HS^-, HPO_4^{2-}, organic acids) are also present in solution, their concentrations must be determined independently and subtracted from the alkalinity value before the DIC can be calculated.

If CO_2 degasing of the water sample is allowed and calcite precipitates in the sample bottle, the alkalinity will not be a conservative parameter.[7] The carbonate in the precipitated calcite represents some of the original alkalinity of the water sample, and a titration of the solution phase will result in an alkalinity value lower than the true aquifer value. For example, a dilute groundwater sample in equilibrium with calcite at a P_{CO_2} of 0.01 atm will have an alkalinity of about 160 mg/L (as $CaCO_3$) and a pH of 7.3. When this sample equilibrates with atmospheric air (P_{CO_2} = 0.0003 atm), CO_2 gas will be lost from the sample, the pH will increase to 8.3, calcite will precipitate (1.1 millimoles/L), and the alkalinity of the water will drop from 160 to 48 mg/L as $CaCO_3$. The calculated bicarbonate concentration of the water sample will decrease from about 200 to 60 mg/L. For this reason it is best to perform the alkalinity measurement on the sample in the field before loss of CO_2 gas. If this is not possible, then a separate water sample should be taken for the alkalinity determination and the entire volume of the sample should be titrated; that is, no subsamples of the water should be used for the alkalinity analysis. In this way any calcite, or other carbonate minerals that have precipitated in the sample bottle will redissolve during acid titration, and their carbonate content will be included in the alkalinity measurement. The loss of CO_2 gas itself from the water sample does not affect the alkalinity because the shift in relative concentration of the bicarbonate/carbonate species with pH increase compensates the alkalinity of the sample for the loss of total inorganic carbon as CO_2 is lost to the air.

The most common unit used to report alkalinity is mg/L as $CaCO_3$. This unit is a source of confusion when converting to bicarbonate and carbonate concentrations because each mole of $CaCO_3$ can neutralize two moles of hydrogen. As a consequence, at pH values less than about 9.3, where HCO_3^- is the dominant inorganic carbon species providing alkalinity to the water, each mole of alkalinity reported as $CaCO_3$ corresponds to two moles of HCO_3^-. For example, if the alkalinity is reported as 100 mg/L as $CaCO_3$, the calculation of the approximate HCO_3^- concentration in the sample is as follows:

$$100 \text{ mg/L as } CaCO_3 \times 1 \text{ g}/1000 \text{ mg} \times 1 \text{ mole}/100 \times 2 \text{ equiv/mole} = 2 \times 10^{-3} \text{ equiv/L}$$

$$2 \times 10^{-3} \text{ equiv/L} \times 1 \text{ equiv/mole} \times 61 \text{ g } HCO_3^-/\text{mole} \times 1000 \text{ mg/g} = 122 \text{ mg } HCO_3^-/L$$

Ignoring the number of equivalents of CO_3^{2-} per mole would result in a calculated HCO_3^- concentration of half the actual value. At pH values greater than about 9.3, where CO_3^{2-} becomes a larger component of the alkalinity, the error decreases. Equations 1-1 and 1-2 (Chapter 1) can be used to convert a reported alkalinity value in terms of mg/L $CaCO_3$ to the respective concentrations of HCO_3^- and CO_3^{2-}. These equations assume that other species that provide alkalinity to the solution are present in minor concentrations or their alkalinity has been subtracted from the total alkalinity value.

In some cases the alkalinity may be reported in units other than mg/L $CaCO_3$. Other possibilities are mg/L CO_3^{2-}, mg/L HCO_3^-, and milliequivalents/L. If the units of alkalinity are stated, then an appropriate conversion can be made to determine concentrations of bicarbonate and carbonate. Finally, the dissolved inorganic carbon may be reported as one of the inorganic carbon species (such as HCO_3^-) and not as alkalinity. In this case the reported concentration is the sum of all the inorganic carbon species in solution [CO_2(aq), H_2CO_3, HCO_3^-, CO_3^{2-}], and the ion speciation relationships (Chapter 1) must be used to calculate the concentrations of the individual species.

9.6. MAJOR CATIONS AND ANIONS

The concentrations of major dissolved ions in groundwater must be measured to understand the geochemical system because these ions are the primary contributors to solution ionic strength, which impacts the effective concentrations of all dissolved species. Also, the major ions may form strong solution complexes with each other and the trace constituents, thereby affecting their mobility and reactivity. Furthermore, a knowledge of the concentration of the major constituents is necessary to calculate saturation indices and identify reactive minerals in the aquifer.

For geochemical calculations, dissolved concentrations are required. Water samples for analysis should be pumped with positive pressure through a 0.45-micrometer pore size filter before preservation to minimize the potential for suspended particulates to bias the analysis. (Pumping under negative pressure, using a vacuum apparatus, may induce degasing of the water, which can influence some of the parameters as discussed above.) For overall transport calculations, colloidal movement may augment solute transport at particular sites where colloid concentrations approach or exceed dissolved levels.

The dissolved concentrations of Na^+ and K^+ in the water sample are relatively stable, however, Ca^{2+}, and perhaps Mg^{2+}, may precipitate as carbonate minerals during CO_2 degasing of a water sample as discussed above in Sections 9.2 and 9.5. For this reason the cation sample bottle is acidified to a pH of 2 at which point the minerals are soluble and will not precipitate. The typical major anions Cl^- and NO_3^- are not particularly reactive in a water sample; therefore the anion sample bottle is generally only chilled after filtration. Chilling the sample will inhibit biological processes that may produce denitrification (loss of NO_3^-) if enough organic carbon is present in the sample to consume available oxygen and then use the nitrogen in nitrate as an electron acceptor. As discussed in Section 9.5, the bicarbonate and carbonate concentrations are generally not stable and should be measured in the field by alkalinity titration.

Most of the major constituents in groundwater are reported by the laboratory in terms of milligrams per liter of the element (e.g. Na^+, Ca^{2+}, Cl^-); however, nitrate is sometimes reported in terms of NO_3^- as N (nitrate as nitrogen). In this case the concentration of the dominant species of nitrogen in the groundwater, NO_3^-, is reported in terms of the element, N. In units of milligrams per liter, a concentration of 10 mg/L NO_3^- as N is equivalent to 44 mg/L NO_3^- as NO_3^-. The difference factor of 4.4 is simply the ratio of the molecular weight of NO_3^- (44 g/mole) to the atomic weight of N (10 g/mole). This method of reporting solution concentrations may also be used for PO_4^{3-}, which is often reported in terms of P and for SiO_2, which may be reported as Si. To interpret laboratory data for compounds reported in this manner, it is necessary to know not only the reporting units but the species or element used to represent the compound.

9.7. MINOR AND TRACE DISSOLVED CONSTITUENTS

The sampling of minor and trace constituents of groundwater is similar to the process for major cations. Because the elements, particularly iron, may precipitate as the pH in the water sample increases due to CO_2 degasing and because cations and anions may adsorb onto the precipitating solids and the walls of the sampling container, the water sample is usually preserved with acid to a pH of 2. Organic compounds may be protected from biodegradation by acidifying the solution or adding a biocide such as $HgCl_2$.

In some cases the components of a redox couple may be separately preserved so that individual concentrations can be determined in the laboratory. These concentrations can be used with the Nernst Equation to calculate an Eh value that corresponds to the dissolved concentrations of the redox species. For example, Fe(II)/Fe(III) and As(III)/As(V) can be preserved by lowering the pH to 2 with HCl with perhaps additional measures taken such as refrigeration and deoxygenation.[8-11] Nitric acid is not recommended as a preservative for redox-sensitive species because of its possible oxidizing effect on the species. Alternatively, if the dissolved concentration of one redox species of an element is controlled by the solubility of a mineral containing another redox species of that element, then only the dissolved concentration of the single redox species need be measured. The Nernst equation relating the concentration of the dissolved species in equilibrium with the mineral can be used to calculate the Eh. This approach is most commonly used with the iron system where the dominant dissolved iron redox species is Fe(II) and the Fe(III) mineral ferrihydrite controls the iron concentration in solution under oxidizing conditions in the typical pH range of groundwater.[12,13]

9.8. REACTIVE SOLID PHASES

The presence of reactive solid phases in an aquifer will limit the groundwater concentration of some of the components of the minerals and will affect the mobility of inorganic and organic compounds by ion exchange and surface complexation reactions with the solids. The identification of minerals that might be limiting solution concentrations in a particular aquifer is partly made by determining the solid phases that are in equilibrium with the groundwater. This is done by calculating the saturation indices of the minerals using a complete water analysis. Confirmation of the presence of these concentration-limiting minerals in the aquifer requires sampling and analysis of the aquifer solids. A review of methods for soil and rock sampling may be found in Davis et al.[14]

In some cases a reactive mineral such as dolomite, pyrite, or Ca-feldspar may irreversibly dissolve into the groundwater and will always be undersaturated with respect to the solution composition. The amount of these minerals present in the solid phase may also be an important geochemical parameter because these minerals can drive many of the reactions involving the formation of secondary minerals that limit solution concentration. To evaluate the potential importance of adsorption processes on the geochemistry of an aquifer, the type and concentration of inorganic sorbents such as ferrihydrite must be measured and the fraction of organic carbon in the solid phase must also be determined.

The impact of sampling and storage on the stability of aquifer solid phases is mainly a response to changes in temperature, water saturation, and gas-phase composition when the solid is removed from the aquifer and exposed to earth surface conditions. The solubilities of minerals are temperature sensitive with most minerals becoming more soluble as the temperature increases; the major exception is the carbonate minerals that are less soluble as temperature increases. In addition, the rate of most chemical reactions is enhanced by temperature increase. As a consequence, if the temperature of the aquifer solid sample changes, this may lead to some dissolution or precipitation of a reactive mineral, thereby changing its solid-phase concentration. In general, solid samples are stored at a low temperature (4°C) to inhibit biological reactions. The amount of mineral dissolution or precipitation that might accompany a change in aquifer temperature from 10° or 20°C to 4°C is not usually considered to have a significant impact on mineral concentration.

Most solid samples are also protected from drying. If evaporation of the pore water is allowed to occur, then solution concentrations will increase and minerals may become oversaturated. This can lead to precipitation of minerals in the sample that are not naturally occurring in the aquifer, but which become stable at higher solution concentrations. If these sampling anomalies are identified in the solid sample and their presence is used to explain site geochemistry, then erroneous interpretations of water/rock interactions are likely. For this reason the porewater in samples should not be allowed to evaporate, and evaporite minerals identified in an aquifer sample should be greeted with skepticism unless ground-water concentrations are very high.

The greatest potential impact of sampling and storage of the solid phase can come from exposure to atmospheric air. It is extremely difficult to eliminate all contact with air when collecting a solid sample and some air may be trapped in the container with the sample. The likelihood of precipitation of calcite in the sample as the water degases CO_2 has been described above (Sections 9.2 and 9.5). Depending on the initial amount of calcite in the solid phase, the amount added by degasing may or may not be significant. Of perhaps greater importance is the presence of solids in the sample that are stable in an anaerobic aquifer but not stable under aerobic surface conditions. For example, under reducing conditions sulfide minerals are stable, but if oxygen is present, sulfide will oxidize to sulfuric acid and several possible solid-phase byproducts. A sample containing pyrite that is contaminated by oxygen in the air will be extremely reactive. Pyrite and calcite, if present, will dissolve, and ferrihydrite and gypsum can form. The precipitation of ferrihydrite will probably change the adsorption properties of the solid. The overall effect is that oxygen contamination of the sample may change the reactive mineral assemblage from one dominated by pyrite and calcite to an assemblage dominated by ferrihydrite and gypsum with perhaps some residual calcite.

Reactive organic matter in the solid is metastable under reducing conditions because of slow reaction rates; however, if oxygen is introduced into the sample, the creation of aerobic conditions will stimulate bacterial growth and enhance biodegradation. This will lower concentrations of organic carbon in the solid phase and increase the partial pressure of CO_2 in the sample. The increase in CO_2 may drive other weathering reactions such as the dissolution of carbonate minerals. The oxidation reactions affecting pyrite and organic matter will have an impact on the oxidizing and reducing capacities of the solid phase described in Chapter 1.

To ensure representativeness of the aquifer solid samples, the impact of geochemical processes on the concentrations of solid phases must be minimized by either protecting the sample from the surface environment or analyzing the sample before reactions have produced a significant change in the solid. If minimizing the reactions cannot be achieved, then an estimate of the degree of impact must be made by considering the reactions that have occurred in the solid between the time of sampling and analysis. It may be difficult to differentiate between aquifer water/rock processes and sampling artifacts.

REFERENCES

1. Stumm, W. and Morgan, J.J., *Aquatic Chemistry,* John Wiley, New York, 1981.
2. Becking, L.G.B., Kaplan, I.R., and Moore, D., Limits of the natural environment in terms of pH and oxidation-reduction potentials, *J. Geol.,* 68, 243–284, 1960.
3. Lindberg, R.D. and Runnells, D.D., Ground water redox reactions: an analysis of equilibrium state applied to Eh measurements and geochemical modelling, *Science,* 225, 925–927, 1984.
4. Robbins, G.A., McAninch, B.E., Gavas, F.M., and Ellis, P.M., An evaluation of soil-gas surveying for H2S for locating subsurface hydrocarbon contamination, *GWMR,* 124–132, 1995.
5. Ballestero, T., Herzog, B., Evans, O.D., and Thompson, G., in *Practical Handbook of Ground-Water Monitoring,* Nielsen, D.M., Eds., Lewis, Chelsea, MI, 1991, pp. 97–141.
6. Fritz, S.J., A survey of charge-balance errors on published analyses of potable ground and surface waters, *Ground Water,* 32, 539–546, 1994.
7. F. J. Pearson, J., Fisher, D.W., and Plummer, L.N., Correction of ground-water chemistry and carbon isotopic composition for effects of CO_2 outgassing, *Geochim. Cosmochim. Acta,* 42, 1799–1807, 1978.
8. Cherry, J.A., Shaikh, A.U., Tallman, D.E., and Nicholson, R.V., Arsenic species as an indicator of redox conditions in groundwater, *J. Hydrol.,* 43, 373–392, 1979.

9. Nicholson, R.V., Cherry, J.A., and Reardon, E.J., Migration of contaminants in groundwater at a landfill: a case study. VI. Hydrogeochemistry, *J. Hydrol.*, 63, 131–176, 1983.
10. Aggett, J. and Kriegman, M.R., Preservation of arsenic(III) and arsenic(V) in Samples of sediment interstitial water, *Analyst*, 112, 153–157, 1987.
11. Davis, A. and Ashenberg, D., The aqueous geochemistry of the Berkeley Pit, Butte, Montana, U.S.A., *Appl. Geochem.*, 4, 23–36, 1989.
12. Beaucaire, C. and Toulhoat, P., Redox chemistry of uranium and iron, radium geochemistry, and uranium isotopes in the groundwaters of the Lodeve Basin, Massif Central, France, *Appl. Geochem.*, 2, 417–426, 1987.
13. Deng, Y. and Stumm, W., Reactivity of aquatic iron(III) oxyhydroxides — implications for redox cycling of iron in natural waters, *Appl. Geochem.*, 9, 23–36, 1994.
14. Davis, H.E., Jehn, J., and Smith, S., Monitoring well drilling, soil sampling, rock coring, and borehole logging. In *Practical Handbook of Ground-Water Monitoring*, Nielsen, D.M., Eds., Lewis, Chelsea, MI, 1991, pp. 195–237.

PART III.
APPLICATIONS
OF GEOCHEMISTRY TO SPECIFIC
TYPES OF CONTAMINANTS
AND CONTAMINATED
ENVIRONMENTS

Part I of this book contains a description of the basic geochemical processes that influence the composition and concentration of groundwater and aquifer solid phases, and Part II discussed the general application of geochemistry to geochemical modeling, contaminant movement, aquifer restoration, and the design of sampling programs. Part III focuses on particular environments such as landfills (Chapter 10) and acidic sites (Chapter 12) or types of contaminants such as metals (Chapter 11) and organic compounds (Chapter 13) where chemical interactions play an important role in the movement and remediation of contaminants.

10 PRACTICAL APPLICATIONS: LANDFILLS

Landfills are the Pandora's Box of potential contamination. They may be of any size from a few acres to hundreds of acres containing any and every type of material, including municipal, medical, industrial, construction, and radioactive waste. Older landfills are likely to contain a wider variety of wastes, to be unlined, and to have waste emplaced either near the water table or actually contacting the groundwater at higher water levels. One common feature of most landfills is that they contain large amounts of readily degradable organic matter. The presence of this mass of reactive, organic material concentrated in a relatively small area produces a localized geochemical environment that influences contaminant migration from these sites.

10.1. GEOCHEMICAL ENVIRONMENT

Figure 10-1 shows the important geochemical environments associated with a typical landfill. Upgradient of the landfill, the shallow aquifer is oxidizing because of the presence of naturally occurring dissolved oxygen in the groundwater. The stable iron and manganese solids are the oxides and oxyhydroxides of Fe(III) and Mn(IV). Because these minerals are not very soluble under oxidizing conditions at normal groundwater pH values (6 to 8.5), the dissolved concentrations of iron and manganese are generally less than 0.1 mg/L in the natural environment. The presence of the landfill disrupts the established geochemical equilibrium between dissolved concentrations and solid/gas phase composition. Rainfall or snowmelt entering the landfill is initially oxidizing because of its contact with oxygen in atmospheric air. However, as the solution percolates through the landfill it reacts with degradable organic matter, and dissolved oxygen is consumed according to the following simplified reaction:

$$CH_2O + O_2 \rightarrow CO_2 + H_2O \tag{10-1}$$

In this reaction carbohydrate (CH_2O) represents all the different potential types of organic matter in the landfill. Other organic compounds may utilize more or less oxygen, but the final result will be oxidation to an inorganic form of carbon; represented in Equation 10-1 as carbon dioxide (CO_2). This reaction shows the consumption of organic matter and dissolved oxygen to produce CO_2. In the case of a young landfill, there will be insufficient oxygen available to the percolating water to consume all of the available reactive organic matter, and the dissolved oxygen concentration will decrease to low levels, producing a leachate with a low redox potential and the capacity to reduce oxidized species that it encounters.

The production of CO_2 by oxidation of organic matter also lowers the pH of the groundwater. As discussed in Section 9.2, high CO_2 concentrations increase the amount of carbonic acid in solution, creating a more acidic solution. Figure 10-2 shows the effect of the municipal waste landfill at Ft. Lewis, Washington, on the groundwater pH and Eh of the shallow aquifer. Upgradient of the landfill, conditions are oxidizing (Eh = +400 mV)

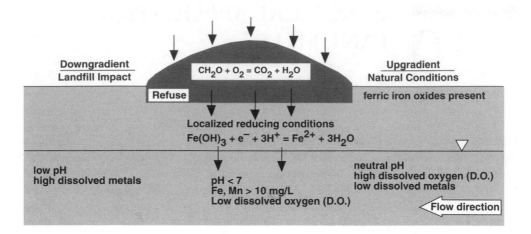

Figure 10-1 Redox reactions — landfill environment.

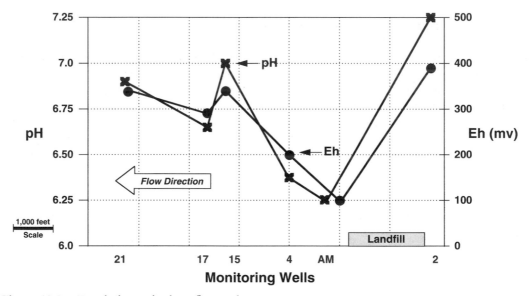

Figure 10-2 pH and Eh trends along flow path.

with a near neutral pH (7.25). Immediately downgradient of the landfill, the Eh has dropped to +100 mV and the pH has been lowered to 6.25. As the landfill-impacted groundwater mixes with fresh recharge water and unimpacted groundwater further downgradient of the landfill, the pH and Eh increase slowly. At the Borden landfill in Ontario (Canada) where depth-discrete samples were collected, it was found that oxygenation of the plume was not very apparent at distances as great as 600 meters downgradient from the landfill.[1] It was surmised that the commonly reported occurrence of oxidizing redox potentials down-gradient of landfills may be partially due to the use of monitoring wells screened within the plume and uncontaminated water above the plume. The resulting mixing of these two waters during sampling may create the apparent oxidizing condition. The monitoring wells at the Fort Lewis landfill were screened over a depth interval of 10 feet near the water table, and the pH/Eh results may partially reflect mixing in the well with fresh recharge water on top of the plume.

When reducing leachate from a landfill contacts manganese and iron oxyhydroxide minerals in the subsoil, the minerals will dissolve because they are more soluble in a reducing environment than in the natural oxidizing environment (Figure 10-1). Because Mn(IV) is reduced at a higher Eh than Fe(III), the oxidized manganese minerals will dissolve before the iron minerals as reducing conditions develop beneath the landfill. However, landfills usually produce such low Eh conditions (<–50 mV) that both the manganese and iron oxide and hydroxide minerals dissolve appreciably. This results in elevated concentrations (>10 mg/L) of dissolved manganese and iron beneath and downgradient of the landfill. Figure 10-3 shows the dissolved concentrations of these two metals in monitoring wells adjacent to the Fort Lewis landfill. Compared with the pH and Eh effects of the landfill on the groundwater (Figure 10-2), the dissolved metal concentrations achieve background levels a shorter distance downgradient of the landfill. This may occur because the Eh of the environment does not have to return to the background value to create a condition in which the manganese and iron solids are relatively insoluble. As discussed in the previous paragraph, some of the lowering of metals concentrations may be due to the use of monitoring wells screened in the plume and in uncontaminated fresh water above the plume.

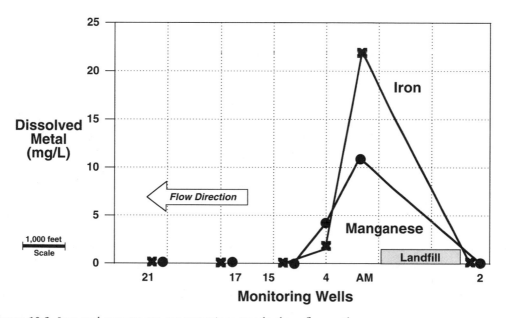

Figure 10-3 Iron and manganese concentrations trends along flow path.

The reducing environment created by the landfill can also affect the concentrations of nitrogen and sulfur species in groundwater. The stable redox species of nitrogen under oxidizing conditions is nitrate (NO_3^-) in which nitrogen is present as N(V). After the available oxygen and Mn(IV) have been reduced in the oxidation-reduction reactions with the landfill organic matter, N(V) in NO_3^- is reduced to NO_2^- [N(III)], N_2 [N(0)] and, ultimately, NH_4^+ [N(-III)]. In a similar manner, but under even greater reducing conditions, sulfate { SO_4^{2-} [S(VI)]} may be reduced to sulfide {S^{2-} [S(-II)]}. The reduction of sulfate to sulfide usually requires the presence of sulfate-reducing bacteria to facilitate the process.

The decomposition of organic matter in a landfill also commonly produces methane by the following fermentation reaction:

$$2CH_2O \rightarrow CH_4 + CO_2 \tag{10-2}$$

In this reaction carbon atoms in the zero valence state [C(0)] in CH_2O have been reduced to C(-IV) in CH_4 and oxidized to C(IV) in CO_2. This reaction occurs under very reducing conditions where other electron acceptors such as oxygen, Fe(III), and N(V) are

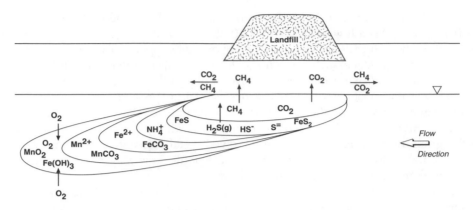

Figure 10-4 Occurrence of dominant redox species and reactive minerals downgradient of landfills.

not available. It also requires the presence of methanogeniic bacteria. In addition to generating methane gas in the landfill, the methane fermentation reaction also produces CO_2, which will lower the pH and add to the dissolved inorganic carbon concentration of the leachate.

Figure 10-4 shows the occurrence of the dominant redox-sensitive species along a flow path from a hypothetical landfill. This distribution correlates with the redox zones from lowest to highest redox potential, except for the location of the NH_4^+ plume. The lowest Eh produced by the landfill occurs in the subsoil closest to the landfill and can be characterized by the presence of methane gas. Methanogenesis also produces carbon dioxide gas. The high concentrations of these two gases in the groundwater may lead to degasing to the vadose zone and vapor-phase transport large distances away from the landfill. Under slightly less reducing conditions in the presence of sulfur and sulfur-reducing bacteria, S(-II) will be the dominant redox-sensitive sulfur species. Depending on the pH, S(-II) will be distributed among the species $H_2S(gas)$, $H_2S(aq)$, HS^-, and S^{2-}. Metal sulfide minerals may form in this zone, particularly FeS and FeS_2. Because the concentration of iron usually exceeds the concentration of S^{2-}, the formation of the iron sulfide minerals will limit the dissolved concentration of sulfide rather than iron.

In terms of the theoretical distribution of redox species, ferrous iron should dominate the next downgradient zone from the landfill; however, NH_4^+, formed from the reduction of NO_3^-, usually occupies the next zone. This occurs because ammonium is strongly attracted to cation exchange sites and its mobility is retarded compared with that of the metals Fe^{2+} and Mn^{2+}. For this reason Figure 10-4 is not a simple representation of redox intensity because it also includes the effect of retardation of contaminants. Ammonium is the only constituent shown on the figure that is both redox sensitive and significantly retarded by exchange reactions.

Ferrous iron, Fe^{2+}, dominates the next zone. It forms from the dissolution of ferrihydrite, and although it does participate in cation exchange reactions, the typical large amount of ferrous iron produced overwhelms the exchange capacity of most systems to significantly retard its movement. In the presence of high ferrous iron and carbonate concentrations, siderite ($FeCO_3$) may form. The precipitation of this mineral usually limits the upper concentration of iron because carbonate is present in excess of the iron concentration. High dissolved levels of manganese may also be present in the iron zone; however, as conditions become more oxidizing the iron will reprecipitate as ferrihydrite [$Fe(OH)_3$], leaving Mn(II) as the dominant, dissolved redox-sensitive species. The manganese carbonate mineral rhodochrosite ($MnCO_3$) may form in the zones where manganese concentration is elevated. Finally, downgradient of the landfill where the plume has sufficiently mixed with fresh water or been subject to the diffusion of gases from the soil vapor, dissolved oxygen will be present at levels greater than 1 mg/L and the system will be oxic again.

The types of reactions described above have been documented by a number of investigators at landfills and other environments with readily degradable organic matter.[2-6] Environments similar to landfills that may create localized reducing conditions include wetlands, leaking underground storage tanks or sewer pipes, organic-rich soil zones, and sewer outfalls. If reducing conditions are naturally occurring and have been established for a long period, then the impact of the reducing environment will not be as significant. For example, in a wetland environment, high dissolved concentrations of iron and manganese may not be present because the oxidized minerals of these metals have been dissolved and no longer exist in the reducing environment. Large anomalies in groundwater composition are generally associated with an induced disequilibrium (such as a landfill or leaking pipe) that disturbs the natural system.

10.2. SAMPLING CONSIDERATIONS

In designing a sampling program to characterize an aquifer impacted by a landfill, contaminants present in the landfill must be considered, as well as the constituents of the aquifer that are mobilized by interactions between the leachate from the landfill and the aquifer solid phases.

The typical major cations in groundwater (Ca^{2+}, Mg^{2+}, Na^+, and K^+) will also be important constituents in an aquifer impacted by a landfill; however, their concentration ratios may be atypical and other cations may also be present at high concentrations. Fritz et al.[7] reported a high concentration level for sodium at a landfill, which they attributed to the presence of road salt in the waste. Because of the reducing environment of the landfill described above, dissolved iron, manganese, and ammonium levels are typically much higher than normal, and their concentrations may exceed the levels of the normal major cations in groundwater. At the Army Creek landfill in Delaware, Baedecker and Apgar[8] found that Fe^{2+} and NH_4^+ accounted for 14 to 50% of the total cations in the anaerobic zone.

Bicarbonate is often the major anion in groundwater associated with landfills because of the production of dissolved inorganic carbon from the oxidation of the organic matter, however, if the bicarbonate concentration is calculated from an alkalinity value, then it is necessary to subtract noncarbonate alkalinity from this value before calculating the inorganic carbon concentration. A significant portion, if not most, of the alkalinity in groundwater near a landfill may be due to the presence of titratable organic anions.[5,8] The concentrations of these organic compounds (that include acetic, oxalic, phenol, benzoic, and tannic acids) must be determined by separate analysis. Higher than normal shallow groundwater concentrations of chloride and sulfate also have been reported for a landfill because of the presence of road salt and gypsum plasterboard ($CaSO_4 \cdot 2H_2O$) in the waste.[1,7] Groundwater nitrate concentrations may be lower than normal because of denitrification accompanying bacterial growth and reducing conditions.

The low pH/high P_{CO_2} condition of groundwater affected by a landfill will create sampling challenges because the CO_2 will degas on contact with atmospheric air, and the pH of the sample will rise (Section 9.2). A landfill setting is one of the most conducive environments for creating disequilibrium conditions between groundwater in an aquifer and a water sample exposed to atmospheric air. Not only will the pH of the water rise as it attempts to equilibrate with CO_2 in the air, but the Eh will probably increase as oxygen from the air dissolves into the water. For these reasons the pH and Eh at landfills need to be measured before the water has been exposed to atmospheric conditions.

Plumb and Pitchford[9] found that the best indicator of organic contamination emanating from a landfill was the presence of the volatile organic compounds dissolved in groundwater. However, as the contaminated water moves away from the landfill and becomes reoxidized, the nonchlorinated compounds may degrade, while the chlorinated volatile compounds appear to be more stable in an oxidizing environment.[8] Eckel et al.[10] report on the presence of pentobarbital, an addictive sedative, in anoxic groundwater 21 years after disposal of the supposedly unstable drug in a landfill. Although pentobarbital is unstable under oxidizing conditions, it apparently can be stable under anoxic conditions. Other compounds not generally considered persistent in the environment may also be stable in the localized reducing environment near landfills and should be considered in the development of sampling plans when it is known that particular types of material, for example medical waste, has been disposed.

10.3. GEOCHEMICAL REACTION MODELING

The development of a conceptual model of the important geochemical reactions in an environment requires information on the reactive solids and gases in the system and the compositions of solutions (rain water, surface water, leachate, and groundwater) that may mix in varying ratios along the flow path. In the case of a landfill, the leachate composition is important, as well as the upgradient and downgradient groundwater. A conservative solution parameter such as the chloride concentration can be used to calculate mixing ratios.[8] In addition to mixing of water types, the other major processes that affect the water composition at a landfill are oxidation of organic matter, mineral dissolution/precipitation, and adsorption/desorption processes. Important consequences of reactions involving organic matter are the consumption of dissolved oxygen, generation of CO_2 and CH_4 gases, and production of organic acids. In developing a model for a landfill-type environment that explains the normally high levels of dissolved inorganic carbon (DIC), it is important to consider the potential sources of DIC from organic matter oxidation, fermentation reactions, and dissolution of carbonate minerals.[6]

A wide variety of reactive minerals should probably be included in a landfill geochemical model. Ferrihydrite ($Fe[OH]_3$) and manganese dioxide (MnO_2) will likely dissolve beneath the landfill and may reprecipitate downgradient of the landfill as the groundwater reoxidizes. Depending on the initial concentration of these metal oxides in the subsoil and the amount of leachate generated, it is possible that all of the original MnO_2 and $Fe(OH)_3$ will dissolve beneath the landfill. If this is the case, the model will need to take into account the limited availability of these solids in the aquifer system. The pH of the aquifer will be buffered by the presence of carbonate minerals, principally calcite, if they are present in the system.[5] Other carbonate minerals may form in the subsoil beneath the landfill because of the high dissolved inorganic carbon concentration produced by oxidation of the organic matter. These minerals may include siderite ($FeCO_3$), $CuCO_3$, smithsonite ($ZnCO_3$), and cerrusite ($PbCO_3$).[3] Under very highly reducing conditions (Eh < −50 mV) in the presence of dissolved sulfide, it is likely that the iron sulfide minerals mackinawite (FeS) and pyrite (FeS_2) will form. It is anticipated that the sulfide minerals will only form and affect groundwater composition in the most reducing zone within or immediately adjacent to the landfill.

The dissolution of strong adsorbents such as ferrihydrite and MnO_2 beneath the landfill will release adsorbed species to the groundwater. These species will likely include anions (of As, Se, and Cr) and cations (Pb, Cu, and Zn). To explain the increase in concentrations of these trace elements in groundwater, the model for the system must include the initial adsorbed concentrations of these elements and release the appropriate amounts to the groundwater as the adsorbent dissolves. Because adsorption affinity is pH and Eh dependent, the model should include a method to account for the effects on adsorbed concentrations due to the low pH and Eh beneath the landfill. In addition to specific adsorption onto metal oxyhydroxides, cation exchange with clays may be an important process at landfills because the localized high concentrations of NH_4^+, Fe^{2+}, and Mn^{2+} compete for the exchange site with the typical major cations. Baedecker and Apgar[2] concluded that a major source of dissolved Ca^{2+}, Mg^{2+}, Na^+, and K^+ in the groundwater at the Army Creek landfill could have been competition for the exchange sites by NH_4^+ in the landfill leachate.

The commonly available geochemical modeling codes (Chapter 5) have been used for a variety of purposes for landfill investigations. Baedecker and Apgar[2] used the BALANCE code to model the mixing process and mass transfer reactions due to mineral precipitation and dissolution. Herczeg et al.[6] used WATEQ4F to calculate dissolved inorganic carbon and CO_2 gas concentrations from measured alkalinity and pH values. Mixing ratios and mineral saturation indices were calculated for a landfill environment by Mirecki and Parks[11] with the PHREEQE code.

10.4. CONTAMINANT TRANSPORT

The contaminants present at a landfill cover the spectrum of organic and inorganic compounds. The organic contaminants are derived from the waste material itself; however, many of the inorganic compounds may come from interactions of landfill leachate with subsurface material. Within the landfill proper and in the leachate moving through the unsaturated zone beneath the landfill, oxidizable organic compounds may be degraded to other organic compounds and ultimately to inorganic carbon compounds. The nonchlorinated organic compounds appear to be most susceptible to oxidation.[2] Complete oxidation of these compounds is rarely achieved, probably due to a lack of sufficient oxygen, and some of these compounds (particularly benzene and toluene from waste petroleum products) are commonly detected in groundwater downgradient of landfills. Reinhard and Goodman[12] attributed the selective removal of xylenes compared with benzene and toluene in a landfill plume to possible biotransformation.

Under anaerobic conditions, methanogenesis complements organic compound oxidation in the degradation and removal of dissolved organic carbon. The chlorinated organic compounds appear to degrade more readily in anoxic conditions, which exist in the aquifer beneath typical landfills.[13] However, anaerobic degradation is a relatively slow process compared with aerobic degradation, and chlorinated compounds can move out of this anaerobic environment into the more oxic environment downgradient of the landfill before being completely degraded. Downgradient of the landfill, the chlorinated compounds are relatively stable and mobile. In addition to biodegradation, the major attenuation mechanism for organic compounds is adsorption onto naturally occurring organic solids in the aquifer. The degree of attenuation is a function of the adsorption affinity for each organic compound (Koc) and the concentration of solid phase organic carbon present in the aquifer. If the adsorption capacity of the system is exceeded due to a large input of contamination, the contaminants will be mobile.[13]

Inorganic compounds, either leaching from a landfill or mobilized by its presence, may include typical major groundwater cations and anions (Ca^{2+}, Mg^{2+}, Na^+, K^+, HCO_3^-, Cl^-, SO_4^{2-}), atypical major species (NH_4^+, Fe, Mn, B), and elevated concentrations of trace elements (Cu, Cd, Pb, Zn, and As).[14-17] The major cations are present in groundwater at elevated levels partially because of cation exchange reactions driven by the high levels of Fe, Mn and NH_4^+ near the landfill. Away from the landfill, as the concentrations of Fe, Mn, and NH_4^+ decrease, the typical major cations will be preferred on the exchange sites; however, they also currently reside on the exchange sites. Therefore cation exchange should not lower their overall solution concentration. As the pH of the groundwater increases downgradient from the landfill due to loss of CO_2 gas to the vadose zone and mixing with water in equilibrium with lower CO_2 gas pressures, magnesian calcite may precipitate and lower calcium and magnesium concentrations in the groundwater. Potassium may be incorporated in the structure of the clay mineral illite and also may decrease in concentration.

The dissolved inorganic carbon concentration of the groundwater should decrease as the groundwater flows away from the landfill as a result of degasing of CO_2 and the possible precipitation of carbonate minerals. The dissolved concentrations of Cl^- and SO_4^{2-} will probably not be affected by geochemical processes, but will decrease as the leachate-impacted groundwater mixes with both fresher water percolating through the vadose zone and unimpacted groundwater. The groundwater ammonium (NH_4^+) concentration will be principally affected by cation exchange reactions. In the reducing environment adjacent to the landfill, the dissolved concentration of iron may be limited by the formation of siderite, $FeCO_3$.[1] Downgradient of the landfill, dissolved iron and manganese concentrations typically return to background levels as the groundwater becomes reoxidized and the oxyhydroxide minerals of these elements become stable again. Boron is a fairly conservative constituent of the groundwater. It does not participate in mineral precipitation or adsorption reactions and is not particularly volatile. For these reasons elevated boron concentrations are expected to persist downgradient of landfills.

In some cases the trace metals Cu, Cd, Pb, and Zn are found in concentrations greater than 1 mg/L in groundwater impacted by landfills. These metals are not normally mobile at high concentrations because they form relatively insoluble carbonate, oxide, or sulfide minerals in either oxidizing or reducing aquifer environments. These metals can be mobile under oxidizing conditions if the pH is low enough that their carbonate minerals are soluble or under reducing conditions if sulfide is not present at a sufficient concentration to limit the metal concentration to a low level. A second process that may enhance the mobility of trace metals is complexation with organic compounds in the leachate. However, Barker et al.[13] did not find this reaction to be an important process for trace metals at the North Bay landfill. They attributed the lack of organic complexation of the trace metals to strong competition for the organic ligands by the major cations in the groundwater.

The mobility of contaminant species such as As and Se that are likely present only at trace concentrations (<1 mg/L) near landfills will be affected primarily by adsorption/desorption processes. Because the anionic form of these contaminants will dominate in the groundwater, the presence of a strong anion adsorber such as ferrihydrite will have the greatest impact on their movement. If ferrihydrite is not present in the aquifer, then the clay minerals will provide the most important exchange sites. It has been shown that pH has a large effect on the amounts of As(V), As(III), and Se(IV) adsorbed from landfill leachate onto clay minerals.[15] Under alkaline conditions, As and Se are not strongly adsorbed onto clay mineral and thus will be mobile in this environment if clay minerals are the only solids present to retard their movement.

The mobilities of the various contaminants at a site are highly dependent on site-specific conditions such as the capacities of the system for neutralization and oxidation. A landfill built on a wetland with a naturally low oxidizing capacity will enable organic compounds in the leachate to travel greater distances in the aquifer than a landfill situated in an oxidizing environment. The reducing zone created by the landfill for the wetland case may exist at some distance from the landfill proper where the leachate/groundwater either discharges at the surface or mixes with naturally oxygenated groundwater. The mobility of Fe and Mn will also be enhanced if the reducing zone extends away from the landfill.

10.5. REMEDIATION

Removing contamination from a groundwater plume emanating from a landfill is complicated by the presence of a wide variety of dissolved contaminants and the divergent conditions under which they can be degraded or mobilized. An *in situ* treatment method that facilitates remediation of one type of contaminant may have little effect on another type. For example, aliphatic and aromatic nonchlorinated organic compounds may be degraded under oxidizing conditions, whereas it appears that chlorinated organic compounds are more susceptible to degradation under anoxic conditions. Furthermore, simple pump-and-treat methods have had little success at landfill sites because of the variety of processes such as diffusion and slow desorption that have limited the rate of removal of contaminants from the aquifer.

Remediation at landfills requires a multiphase approach to address the entire suite of contaminants.

The primary inorganic contaminants associated with landfills are Fe, Mn, and NH_4^+. Each of these contaminants were produced by the reducing conditions generated in the landfill and in the subsurface beneath the landfill. The reestablishment of oxidizing conditions should effectively lower the dissolved concentrations of Fe and Mn to below a level of concern because of the low solubility of their minerals under oxidizing, near-neutral pH conditions. Precipitation of Fe and Mn oxyhydroxides and oxides in the aquifer may also reduce concentrations of trace elements (e.g., As, Se, Cr, and Pb) that may be of concern at particular landfill sites. If natural mixing of the groundwater due to dispersion is not sufficient to generate oxidizing conditions, then oxygen can be sparged into the aquifer via injection wells situated across the flow path downgradient of the aquifer.

The removal of ammonium, NH_4^+, from an aquifer is more difficult than the metals. Ammonium is relatively strongly held on the clay mineral exchange sites; consequently, simple pumping of contaminated water is not a very efficient method for removing this ion. Increased efficiency in the removal of NH_4^+ has been achieved by circulating a divalent cation such as Ca^{2+} and/or Mg^{2+} through the zone of NH_4^+ contamination to compete for the exchange sites and enhance NH_4^+ removal.[18] A second method of removing ammonium is the nitrification process in which NH_4^+ is oxidized with oxygen to nitrite (NO_2^-) and nitrate (NO_3^-). The oxidized forms of nitrogen are anionic and more mobile than NH_4^+; thus they will be easier to remove from the aquifer. Furthermore, some of the nitrogen may be degssed from the aquifer to the unsaturated zone by the denitrification process that produces the gases N_2 and N_2O. Because it is anticipated that the NH_4^+ plume will be adjacent to the Fe plume, *in situ* oxidation for the metals contamination can be designed to enhance the removal of ammonium as well.

The most common method of removing volatile organic compounds from groundwater is by air stripping. This involves contacting the contaminated water with air, thereby allowing the dissolved volatile compounds to transfer from the groundwater to the vapor phase and lowering solution concentrations. Air stripping may take place in a surface tower as part of a pump-and-treat operation or it may be performed *in situ* within a well. In either case, movement of the organic contaminant to the location where stripping occurs may be retarded by adsorption onto organic matter in the aquifer and this will extend the time required to achieve aquifer restoration.

An important design consideration for the air stripping process is the potential for mineral precipitation as a byproduct of groundwater degasing. The minerals most susceptible to precipitation are calcite and iron and manganese oxyhydroxides. The presence of a landfill increases the CO_2 gas partial pressures in shallow groundwater and the concentrations of the components of calcite, Ca^{2+} and dissolved inorganic carbon. Table 10-1 provides data on the concentrations of calcium and dissolved inorganic carbon (as represented by bicarbonate) in an upgradient and downgradient well at the Fort Lewis (Washington) landfill. The concentrations of these two components increase by almost an order of magnitude due to the influence of the landfill. Under aquifer conditions, calcite is undersaturated because of the low pH of the leachate-impacted groundwater; however, the P_{CO2} of the contaminated aquifer is 0.44 atm, which is over three orders of magnitude greater than the atmospheric CO_2 concentration.

Table 10-2 shows the simulated result of air stripping the contaminated water downgradient of the Fort Lewis landfill. The water is allowed to equilibrate with the atmospheric P_{CO2} (0.0003 atm) value and to become oxidized because of oxygen in the air contacting the water. In the simulation, calcite and ferrihydrite [$Fe(OH)_3$] are allowed to precipitate if they become oversaturated in the conditions of the air stripper. It can be seen that 410 mg of calcite and 42 mg of ferrihydrite precipitate per liter of water treated in the air stripper.

TABLE 10-1.

CALCITE MINERAL EQUILIBRIUM DATA, FORT LEWIS LANDFILL		
Parameter	Upgradient well	Downgradient Well
pH	7.14	6.14
Ca (mg/L)	17	164
HCO_3^- (mg/L)	92	820
P_{CO_2} (atm)	0.005	0.44
Calcite saturation index	−1.2	−0.48

TABLE 10-2.

SIMULATED AIR STRIPPING OF FORT LEWIS CONTAMINATED GROUNDWATER		
Parameter	Contaminated water	Air-equilibrated water
pH	6.14	8.8
Eh (mV)	109	400
P_{CO_2} (atm)	0.44	0.0003
Calcite Saturation Index	−0.48	0
Ferrihydrite Saturation Index	−1.58	0
Calcite Precipitated (mg/L)	—	410
Ferrihydrite Precipitated (mg/L)	—	42

Using the molar volumes of calcite (36.9 cm³) and ferrihydrite (30.6 cm³) and a flow rate of 100 gpm through the air stripper, this would equate to about 82,500 cm³ (82.5 L) of calcite and 6600 cm³ (6.6 L) of ferrihydrite precipitating in the void spaces of the air-stripping tower per day. As discussed in Section 6.3, this maintenance issue is usually addressed in a surface treatment plant by adding an acid to increase the solubility of these minerals under the surface conditions. For an *in situ* treatment system, adding CO_2 gas to the air used for stripping has proven effective and economical.[19]

In situ biodegradation is a remediation method that can be effective for both volatile and nonvolatile organic contamination. For those compounds that best degrade under oxic conditions, oxygen can be added in the form of air, oxygen gas, or hydrogen peroxide. In this case delivery of oxygen throughout the area of contamination is the complicating factor and not the removal of an attenuated contaminant. Oxygen is typically added by injection wells that may intersect the leading edge of the plume or wells located throughout the plume's areal extent.[20] The in-well air-stripping system also achieves some remediation benefit from oxidation because the water circulation pattern developed by the injection of air into the well also serves to oxygenate the aquifer. The funnel-and-gate remediation system discussed in Chapter 8 can also be used to create localized oxidizing conditions in an aquifer by using an oxygen-releasing solid in the reactive gate material.[21]

For compounds that are more susceptible to degradation under a reducing environment, conditions must be enhanced for either bacterial- or chemical-reducing processes. For the chlorinated compounds present in groundwater at the part-per-billion level, it has been shown that they can be effectively cometabolized and biodegraded if another food source (such as methane or toluene) is present to stimulate bacterial growth.[22,23] Conditions remain anoxic and conducive to degradation of the chlorinated compounds because of the amount of organic food source added to the system. Chemical reduction of the chlorinated compounds such as trichloroethene and tetrachloroethene has been shown in the laboratory to be effective in the presence of a strong reductant such as zero-valent iron.[24] Starr and Cherry[25] have recommended the use of zero-valent iron in a funnel-and-gate system to treat halogenated organic contaminants. They also envision a possible series of gates to treat different types of compounds. These gates might consist of (1) metallic sand to treat contaminants susceptible to anoxic degradation, (2) oxygen-releasing compounds to enhance aerobic biodegradation, and (3) neutralizing minerals to immobilize metals.

REFERENCES

1. Nicholson, R.V., Cherry, J.A., and Reardon, E.J., Migration of contaminants in groundwater at a landfill: a case study. VI. Hydrogeochemistry, *J. Hydrol.*, 63, 131–176, 1983.
2. Baedecker, M.J. and Back, W., Hydrogeological and Chemical Reactions at a Landfill, *Ground Water*, 17, 429–437, 1984..
3. Yanful, E.K., Nesbitt, H.W., and Quigley, R.M., Heavy metal migration at a landfill site, Sarnia, Ontario, Canada. I. Thermodynamic assessment and chemical interpretations, *Appl. Geochem.*, 3, 523–533, 1988.
4. Farquhar, G.J., Leachate: production and characterization, *Can J. Civ. Eng.*, 16, 317–325, 1989.
5. Kehew, A.E. and Passero, R.N., pH and redox buffering mechanisms in a glacial drift aquifer contaminated by landfill leachate, *Ground Water*, 28, 728–737, 1990.
6. Herczeg, A.L., Richardson, S.B., and Dillon, P.J., Importance of methanogenesis for organic carbon mineralisation in groundwater contaminated by liquid effluent, South Australia, *Appl. Geochem.*, 6, 533–542, 1991.
7. Fritz, S.J., Bryan, J.D., Harvey, F.E., and Leap, D.I., A geochemical and isotopic approach to delineate landfill leachates in an RCRA study, *Ground Water*, 32, 743–750, 1994.
8. Baedecker, M.J., Cozzarelli, I.M., Eganhouse, R.P., Siegel, D.I., and Bennett, P.C., Crude oil in a shallow sand and gravel aquifer. III. Biogeochemical reactions and mass balance modeling in anoxic groundwater, *Appl. Geochem.*, 8, 569–586, 1993.
9. R. H. Plumb, J. and Pitchford, A.M., *Volatile Organic Scans: Implications for Ground Water Monitoring*, Ground Water Publishing, Proceedings NWWA/API Conference on Petroleum Hydrocarbons and Organic Chemicals in Groundwater—Prevention, Detection, and Restoration, 1985.
10. Eckel, W.P., Ross, B., and Isensee, R.K., Pentobarbital found in ground water, *Ground Water*, 31, 801–804, 1993.
11. Mirecki, J.E. and Parks, W.S., Leachate geochemistry at a municipal landfill, Memphis, TN, *Ground Water*, 32, 390–398, 1994.
12. Reinhard, M., Goodman, N.L., and Barker, J.F., Occurrence and distribution of organic chemicals in two landfill leachate plumes, *Environ. Sci. Technol.*, 18, 953–961, 1984.
13. Barker, J.F., Tessmann, J.S., Plotz, P.E., and Reinhard, M., The organic geochemistry of a sanitary landfill leachate plume, *J. Cont. Hydrol.*, 1, 171–189, 1986.
14. Garland, G.A. and Mosher, D.C., Leachate effects of improper land disposal, *Waste Age*, 42–48, 1975.
15. Frost, R.R. and Griffin, R.A., Effect of pH on adsorption of arsenic and selenium from landfill leachate by clay minerals, *Soil Sci. Soc. Am. J.*, 41, 53–57, 1977.
16. Lu, J.C.S., Eichenberger, B., and Stearns, R.J., *Leachate from Municipal Landfills: Production and Management*, Noyes, NJ, 1985.
17. Bolton, K.A. and Evan, L.J., Elemental composition and speciation of some landfill leachates with particular reference to cadmium, *Water, Air, and Soil Poll.*, 60, 43–53, 1991.
18. Yan, T.H. and Espenscheid, W.F., *Removal of Ammonium Ions from Subterranean Formations by Flushing with Lime Saturated Brines*, Society of Mining Engineers, Denver, 1982.
19. Peterson, S., Personal communication, 1995.
20. Pankow, J.F., Johnson, R.L., and Cherry, J.A., Air sparging in gate wells in cutoff walls and trenches for control of plumes of volatile organic compounds, (VOCs), *Ground Water*, 31, 654–663, 1993.

21. Bianchi-Mosquera, G.C., Allen-King, R.M., and Mackay, D.M., Enhanced degradation of dissolved benzene and toluene using a solid oxygen-releasing compound, *GWMR,* 120-128, 1994.
22. Wilson, J.T. and Wilson, B.H., Biotransformation of trichloroethylene in soil, *Appl. Envir. Microbiol.,* 49, 242–243, 1985.
23. Hopkins, G.D. and McCarty, P.L., Field evaluation of in situ aerobic cometabolism of trichloroethylene and three dichloroethylene isomers using phenol and toluene as the primary substrates, *Environ. Sci. Technol.,* 29, 1628–1637, 1995.
24. Gillham, R.W. and O'Hannesin, S.F., Enhanced degradation of halogenated aliphatics by zero-valent iron, *Ground Water,* 32, 958–967, 1994.
25. Starr, R.C. and Cherry, J.A., In situ remediation of contaminated ground water: the funnel-and-gate system, *Ground Water,* 32, 465–476, 1994.

11 PRACTICAL APPLICATIONS: METALS CONTAMINATION

Most of the elements on the Periodic Table are metals. The metallic elements are classified as alkali metals (Li, Na, K, Rb, Cs, and Fr), alkaline earth metals (Be, Mg, Ca, Sr, Ba, and Ra), transition metals (e.g., V, Cr, Fe, Mo, Hg), and semimetals or metalloids (e.g., B, Si, As). The lanthanide and actinide series of elements are also metals. Metals are elements that give up electrons in their outer orbitals to achieve a more stable electron configuration. In so doing, the elements become positively charged, although as a result of complexation reactions in solution, the dominant dissolved metallic species may be an anion such as chromate CrO_4^{2-}. To facilitate the site characterization process at potentially contaminated sites, the U.S. Environmental Protection Agency has developed a number of lists of metals of interest. Table 11-1 provides three of the lists of metallic elements. Most of these metals represent some human health or environmental risk if concentrations exceed a threshold level that is dependent on the medium, type of receptor, exposure duration, and other factors. Although the metals Ca, Mg, Na, and K are not considered a potential threat to the environment, they are included in the Target Analyte List for groundwater characterization purposes because these metals are typically the major dissolved cations.

This chapter describes the geochemistry of a few metals (Pb, Cr, and As) that are representative of metals contamination in the environment. The sources of contamination for these metals are discussed in the context of the geochemical impact of contact of the source material with the natural environment. Methods of modeling the geochemistry of this interaction are discussed, as well as contaminant migration and remediation of the metal of interest. Appendix A contains a summary table with information on the water/rock interactions that can limit the solution concentrations of a wide variety of metal and nonmetal contaminants in soil and aquifer environments.

11.1. GEOCHEMICAL ENVIRONMENT

When contaminants enter the natural environment, interactions occur between the contaminating source material and the subsurface solution, solid, and gas phases. If the contaminant is reactive and the immobilization capacity of the natural system is sufficient to bind the contaminant to the solid phase or result in degasing of a volatile species of the element, then subsurface movement of the contaminant in an aquifer may be minimal. If the contaminant influx overwhelms the capacity of the system to immobilize the contaminant, then the contaminant will be relatively mobile until a sufficient mass of subsurface material has been contacted to retard or immobilize the contaminant. The following describes typical geochemical responses of the natural environment to Pb, Cr, and As contamination from various sources.

Lead

Sources of lead contamination to the environment include fairly localized impacts such as disposal of lead-acid batteries, lead-based paint wastes in landfills, and tailings from the processing of Pb minerals, principally galena (PbS). More widespread occurrences of lead

TABLE 11-1.

METAL ANALYTE LISTS

Priority pollutants	CLP target analyte list	RCRA metals
Antimony (Sb)	Aluminum (Al)	Arsenic (As)
Arsenic (As)	Antimony (Sb)	Barium (Ba)
Beryllium (Be)	Arsenic (As)	Cadmium (Cd)
Cadmium (Cd)	Barium (Ba)	Chromium (Cr)
Chromium (Cr)	Beryllium (Be)	Lead (Pb)
Copper (Cu)	Cadmium (Cd)	Mercury (Hg)
Lead (Pb)	Calcium (Ca)	Selenium (Se)
Mercury (Hg)	Chromium (Cr)	Silver (Ag)
Nickel (Ni)	Cobalt (Co)	
Selenium (Se)	Copper (Cu)	
Silver (Ag)	Iron (Fe)	
Thallium (Tl)	Lead (Pb)	
Zinc (Zn)	Magnesium (Mg)	
	Manganese (Mn)	
	Mercury (Hg)	
	Nickel (Ni)	
	Potassium (K)	
	Selenium (Se)	
	Silver (Ag)	
	Sodium (Na)	
	Thallium (Tl)	
	Vanadium (V)	
	Zinc (Zn)	

contamination have been associated with aerosols from leaded gasoline and smelters, lead-based paint from buildings, $PbHAsO_4$ used as an insecticide on orchard trees, and sewage sludge applied to agricultural fields.[1] In cases of widespread deposition of lead, the concentrations of lead and other compounds in the contaminating medium (aerosols, paint, insecticide) have been low and application/emission rates have generally been modest, except perhaps very near the source. The soil environment has been able to immobilize the lead in most cases.[2] However, the immobility of lead has resulted in its accumulation in the soil. Because of the toxicity of lead, shallow contaminated soil can pose an ingestion risk.

Lead is immobilized by soil primarily by adsorption and mineral precipitation processes. The dominant dissolved Pb species in dilute water up to a pH of about 7 is Pb^{2+}. Above this pH, $PbCO_3^0$ and $PbOH^+$ are the dominant species.[3] The $PbSO_4^0$ complex is important at SO_4^{2-} concentration greater than about 10^{-3} mol/L (96 mg/L). In the soil environment, where organic matter is relatively abundant, investigators have found that soil humus plays the major role in immobilizing Pb added to the soil.[4,5] Adsorption onto the soil organic matter has been best described using the Langmuir isotherm (Chapter 3). Lead is also strongly adsorbed onto Fe-Mn oxides, and below the soil zone where solid organic carbon is less abundant it is expected that adsorption onto inorganic solids plays a major role in immobilizing lead.

Several relatively insoluble lead minerals may also limit the mobility of lead in soil and aquifer environments. In noncalcareous soils the solution concentration of Pb has been found to be limited by $Pb(OH)_2$, $Pb_3(PO_4)_2$, and $Pb_5(PO_4)_3OH$, and in calcareous soils $PbCO_3$ can be a reactive solid.[6] The solubilities of each of these minerals will be a function of pH and the solution concentrations of the other components (PO_4^{3-} and CO_3^{2-}) in the lead mineral. If the influx of lead contamination to the environment exceeds the available phosphate or carbonate supply to solution, then these minerals will not affectively limit the lead concentration in solution.

When the zone of contamination is localized and lead concentrations in the source are high (such as past practices of disposing of battery acid in unlined pits and paint products in landfills without liners), the amount of lead entering the subsoil may exceed the immobilization capacity of the system, allowing lead to be relatively mobile. Lead may also form organic complexes with the solvents in paint, thereby increasing its mobility. In the case of battery acid, the adsorbents in the subsurface may be dissolved and removed by the acid entering the environment, and most lead minerals are more soluble under acidic conditions. This will enhance the movement of lead. The actual mobility of lead in a particular environment subjected to high levels of contamination will be a function of the adsorption capacity of the various adsorbents (organic matter and metal oxides), the solubility of the lead minerals, and, in the case of an acidic waste stream, the neutralizing capacity of the subsurface material. In most environments lead is expected to be relatively immobile, and groundwater contamination by lead has not been of major concern. However, in a study of 71 North Carolina landfills, Borden and Yanoschak[7] found that groundwater at 18% of the landfills exceeded the lead drinking water standard of 0.05 mg/L. The mean total concentration of lead was 0.078 mg/L with a standard deviation of 0.265. Because only unfiltered water samples were collected, it is possible that the reported concentrations include a portion of lead on particulates that may not be mobile in the aquifer.

Chromium

Chromium is a valuable metal that is used in a wide variety of products and applications such as paint, metal alloys (stainless steel), hard chrome plating, corrosion inhibitors, refractory bricks, printing inks, photographic film, wood preserving, and leather tanning.[8,9] Waste products from these industries can be a source of chromium contamination to the environment. Releases to the atmosphere in the form of particulates generated by electric arc furnaces used in ferrochrome and refractory brick production provide the largest total amount of chromium released by human activity. Disposal of fly-ash from coal combustion provides the largest single input to soils. Another widespread source of chromium to soils is the application of sewage sludge to the land surface as a disposal method. Sludges from tanneries or plating facilities can produce locally high concentrations of chromium. In some locations chromate-bearing slag from the processing of chromite ore has been used as general fill material that may pose a human health or environmental hazard.[10]

At trace concentrations chromium appears to be an important nutrient in the human diet; however, as with other trace elements, it is also toxic at above average concentrations. In cases of very high exposure Cr can be toxic and carcinogenic.[8] Cr(VI) is usually identified as the toxic form of chromium because it occurs as the fairly mobile anion chromate (CrO_4^{2-}), and its minerals are soluble in the natural environment. Relatively high concentrations of Cr(VI) are possible in the environment compared with Cr(III), which forms an insoluble hydroxide mineral, $Cr(OH)_3$. Many industries use the chromate and dichromate ($Cr_2O_7^{2-}$) forms of chromium for such applications as corrosion inhibition because of the strong oxidizing ability of these species. It has been reported that the dust emitted from ferrochrome smelters contains on the order of 40% chromium as Cr(VI), whereas chromium in sewage sludge is predominantly in the Cr(III) valence state because of the reducing environment.[8]

Because of the potential health and environmental impact of Cr(VI), the interaction of Cr(VI) species with the natural environment is of primary concern. Cr(VI) species adsorb onto soil material, although generally less strongly than Cr(III) species. Furthermore, Cr(VI) minerals are more soluble than Cr(III) minerals at most soil and aquifer pH/Eh values. The stable redox state of chromium in the environment is predominantly Cr(III) (Figure 11-1). Therefore if the natural environment has sufficient reducing capacity, it should respond to the impact of Cr(VI) contamination by reducing Cr(VI) to Cr(III) and immobilizing it as the

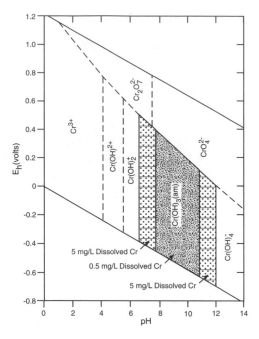

Figure 11-1 Chromium speciation and mineral equilibrium.

solid $Cr(OH)_3$. This redox/precipitation process for immobilizing Cr(VI) is affected primarily by the rate of reduction and the possible maintenance of oxidizing conditions in the subsurface by the presence of Mn(IV) minerals. These factors are discussed further in Section 11.4 on contaminant mobility.

Arsenic
Some of the earliest uses of arsenical compounds were as pesticides and herbicides. Lead arsenate was commonly used to control insect pests in orchards and sodium arsenite was employed to defoliate seed potatoes and clear aquatic weeds. Use of arsenic for these purposes declined about 50% from 1970 to 1980 because of concern for As buildup in soils.[11] The application of arsenic as a preservative for wood in the compounds copper-chromium-arsenate and ammonium-copper-arsenate remains an important use of this semi-metal. Arsenic may also enter the environment through anthropogenic input as an aerosol created by the smelting of copper, lead, zinc, and gold ores and by coal combustion. The particulates in the atmosphere may contaminate downwind soils. Also, arsenic may leach from the ash produced by burning coal.

The dominant redox states of arsenic in the environment are As(III) and As(V). Under slightly oxidizing conditions and above (pe + pH >8), As(V) is the stable redox species, while under more reducing conditions As(III) species dominate. Microorganisms can methylate/demethylate arsenic over a wide range of pH and Eh conditions, producing monomethylarsonic acid $(CH_3AsO[OH]_2)$; dimethylarsinic acid (cacodylic acid) $([CH_3]_2AsO[OH])$; trimethyl arsenic oxide $(CH_3)_3AsO$; trimethylarsine $(CH_3)_3As$, and dimethylarsine $(CH_3)_2AsH$. The reactions producing these compounds depend on the type of organism present and the form of arsenic.

In the pH range of natural waters, the As(V) species are predominantly anionic $(H_2AsO_4^-$ and $HAsO_4^{2-})$ (Figure 11-2). Ferrihydrite has a very strong affinity for these species, as well as a high capacity for adsorption. It will scavenge arsenic from solution. The As(III) species in water are primarily the neutral species $H_3AsO_3^0$ and the anion $H_2AsO_3^-$. The affinity of ferric hydroxide for dissolved As(III) is less than that for As(V).[12] Under oxidizing conditions, the dissolved concentration of arsenic may be limited by the relatively soluble

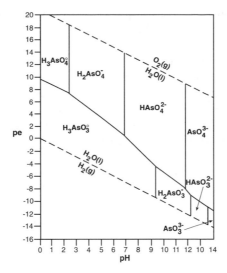

Figure 11-2 Arsenic speciation. (Modified from Rai and Zachara 1984.)

mineral scorodite ($FeAsO_4 \cdot 2H_2O$), while under reducing conditions in the presence of sulfide, the insoluble minerals orpiment (As_2S_3) and realgar (AsS) may limit arsenic solution concentrations.

The acute toxicity of As(III) is substantially greater than that of As(V)[13]; therefore the transformation between redox species is important from the standpoint of potential risk. Methylated arsenic species are apparently less acutely toxic than inorganic species, and the formation of organo-As compounds by organisms may be a mechanism for detoxification of arsenic.

11.2. SAMPLING CONSIDERATIONS

Because metals are commonly included in any characterization of the inorganic composition of an aquifer, the general sampling considerations provided in Chapter 9 are applicable. The mobilities of many of the trace metals are strongly affected by adsorption onto solid-phase organic matter and metal oxides and hydroxides, particularly of iron and manganese. For this reason the analysis of the aquifer solid phase should include a measure of the concentrations of these adsorbents in the aquifer and the amount of trace metals adsorbed on their surfaces. Also, because a large percentage of the suspended matter in a water sample may contain adsorbed metals, care must be exercised in filtering samples with significant particulate matter. If the particulates are not removed by filtering, the adsorbed metals will desorb from the surfaces when the water sample is acidified. If this occurs, it will be difficult to estimate the true dissolved concentrations, which are required for calculations of water/rock interactions. Micropurging of monitoring wells by low-flow pumping has been recommended as a method of collecting groundwater samples without mobilizing particulates by the sampling process.[14]

Reactions that occur in the sample container may also have the effect of reducing dissolved concentrations from the levels present in the aquifer. For example, if the water has a low redox potential and contains high concentrations of dissolved iron and manganese, it will oxidize in the sample container and, prior to acidification, iron and manganese oxides/hydroxides will precipitate in the container. Trace metals will also be removed from solution by coprecipitation/adsorption onto the metallic solids. Filtering the sample after this precipitation has occurred, but prior to acidification, will remove previously dissolved constituents, resulting in low apparent concentrations.

The redox species of arsenic (As[III] and As[V]) can be measured separately in a water sample to calculate the redox potential that corresponds to the dissolved concentrations. The calculated redox potential can then be compared with the platinum-electrode–measured Eh to determine if the arsenic redox species are in equilibrium with the measured Eh. Aggett and Kriegman[15] discuss sample preservation techniques to minimize oxidation or reduction of arsenic in the sample container. They recommend acidification to a pH of 2, refrigeration near 0°C, and deoxygenation as the safest preservation method, although lesser methods may be sufficient for short term (a few weeks) storage. In contrast, Cherry et al.[16] evaluated the oxidation and reduction of As(III) and As(V) in lab trials with redox agents common to natural water such as O_2, H_2S, and Fe. They found that redox reactions for As species occurred at a rate sufficiently slow to allow samples to be collected and analyzed in the laboratory without any special sample preservation methods.

If a sufficient quantity of a contaminating metal is disposed in the environment, saturation will be reached with respect to a reactive mineral. The mineral will precipitate in the subsurface providing a secondary source of the contaminant. In these cases the reactive mineral must be identified and its concentration measured to design an appropriate restoration method and predict the amount of time required to achieve remediation. If the identity of the reactive mineral is not determined, then its solubility under various restoration scenarios cannot be estimated, and if the amount of the mineral and its distribution in the subsurface is not known, then the degree and duration of restoration (amount of reagent to use and water to pump) cannot be quantified.

The presence and concentration of some naturally occurring minerals may also need to be determined because of their possible impact on restoration. Pyrite may be oxidized during restoration, releasing sulfate and producing acidic conditions if oxidizing conditions are generated in an aquifer to enhance remediation of organics. The low pH conditions generated by oxidizing pyrite may inhibit biodegradation and mobilize trace metals. As a consequence, the possible presence of pyrite and other sulfide minerals should be evaluated in an aquifer before oxidizing conditions are generated for restoration purposes. The immobilization of Cr as $Cr(OH)_3$ may be inhibited if the aquifer contains oxidized manganese minerals (e.g., MnO_2), which will oxidize Cr(III) to Cr(VI) and maintain relatively high mobility for chromium.[17] If the remedial technique consists of *in situ* reduction of Cr(VI) to Cr(III), then enough reactant will have to be added to also reduce manganese present as Mn(IV) in the subsurface solids.

11.3. GEOCHEMICAL REACTION MODELING

Simulating water/rock interactions of metals in the environment requires an accurate conceptual model of aqueous complexation, adsorption/desorption, and mineral precipitation/dissolution. In addition to hydroxyl and other inorganic species, the metal may form organic complexes as in the case of methylated arsenic. The metal may occur in more than one redox state (e.g., Cr[III]/Cr[VI] and As[III]/As[V]) in the aquifer. Adsorption of the metal species may vary drastically [As(V) >> As(III)], and the solubility of their minerals in the aquifer environment may differ by orders of magnitude [$Cr(OH)_3$ < soluble < Cr(VI) minerals]. The computer code used to calculate equilibrium/nonequilibrium solution composition must have the ability to integrate all these geochemical processes.

Adsorption of metals can be modeled by surface complexation models such as the diffuse double-layer model that has been incorporated into the MINTEQA2[18] and PHREEQC[19] codes using the intrinsic adsorption constants derived by Dzombak[21] for iron oxide. In addition to these adsorption constants, site-specific data on the concentration of the adsorbent in the aquifer must be obtained, as well as the concentrations of competing solutes. The alternative to using a surface complexation model is to develop adsorption constants for the isotherm that most closely fits the site data. Rai and Zachara[12] have compiled these constants for single-phase substances (e.g., clays and metal oxides) and multiphase soils for many of the metals that may occur as contaminants. This compilation provides an excellent guide to the probable range of adsorption constants, which can be compared to constants developed for a particular site from groundwater and solid-phase concentrations.

Loux et al.[20] used the MINTEQA2 code to simulate water/rock interactions of eight metals (Ba, Be, Cd, Cu, Ni, Pb, Tl, and Zn) added to groundwater collected downgradient of a landfill. Adsorption was modeled using the adsorption constants derived by Dzombak[21] for pure-phase iron oxide. The modeling results were compared with batch experiments in which groundwater, spiked with contaminant concentrations of approximately 3 mg/L for each metal, was contacted for 48 hours with aquifer material. Modeling and batch experiments were conducted over the pH range of 4 to 10. The results showed that, for the particular site conditions and contaminant levels studied, adsorption best described the pH-dependent partitioning for Pb, Zn, and Ni, while the dissolved Cd concentration was best explained by precipitation of $CdCO_3$. The model did not successfully predict solution concentrations of Cu, Ba, Be, or Tl over the pH range of the study. This may be due to control by a mineral not available in the MINTEQA2 database, inaccurate thermodynamic data in the database, or adsorption by a solid phase that does not have the adsorption properties of iron oxide.

The other potential geochemical process to include in a model of metal mobility is mineral equilibrium. The challenge in modeling mineral equilibrium is in selecting reactive minerals that might form in the environment. Compilations of reactive minerals are available in Lindsay[3] and Rai and Zachara,[12] while other references may include minerals specific to a particular environment (such as a landfill or battery disposal site). In the case of Pb, Baltpurvins[22] considered the following minerals as potentially reactive: blixite ($Pb_4[OH]_6Cl_2$), laurionite (Pb[OH]Cl), anglesite ($PbSO_4$), and cerrusite ($PbCO_3$). A number of other studies mention the likelihood of $PbSO_4$ forming from the weathering of sulfide deposits or at industrial sites contaminated with Pb and sulfate.[23,24] Rai and Zachara add to the list of Pb minerals $Pb(OH)_2$, $Pb_3(PO_4)_2$, $Pb_4O(PO_4)_2$, and $Pb_5(PO_4)_3OH$. For reducing conditions with sulfide present, Lindsay[3] shows that galena (PbS) should be stable.

Chromium (VI) minerals are relatively soluble and would only be expected at sites with very high chromate contamination. Under these conditions it appears possible that hashemite ($BaCrO_4$) may form.[25] Palmer and Wittbrodt[26] report the precipitation of chromatite ($CaCrO_4$) under evaporative conditions in a drainage ditch at a chrome-plating facility. James[10] also reports the presence of the sparingly soluble chromate salt $PbCrO_4$ in waste-amended soils. Under typical aquifer pH/Eh conditions it is anticipated that the Cr(III) minerals $Cr(OH)_3$ or $Cr_xFe_{1-x}(OH)_3$ will be stable and relatively insoluble.[27,28]

Where As(V) is the dominant arsenic redox species, the most commonly reported arsenic mineral is scorodite ($FeAsO_4 \cdot 2H_2O$),[24] which appears to be stable only in the narrow pH range of approximately 4 to 5.[29] There is also some evidence that $Pb_3(AsO_4)_2$ and $Mn_3(AsO_4)_2$ may be limiting arsenic concentrations in soil.[30] Under reducing conditions with sulfide present, the minerals orpiment (As_2S_3) and realgar (AsS) may form, thereby limiting the solution concentration of arsenic.[31]

11.4. CONTAMINANT MOBILITY

The mobilities of Pb, Cr, and As in the subsurface environment will be primarily a function of the types of solution complexes formed, the affinity and capacity of the solid-phase adsorption sites for the contaminant, and the solubilities of reactive minerals containing the contaminant.

Lead

The movement of trace levels of lead in the subsurface appears to be strongly affected by adsorption onto a variety of solid-phase surfaces. In a study of five soils from a heavily industrialized region of northwestern Indiana subjected to extensive metals contamination, Miller and McFee[2] found the top 2.5 cm of soil to be highly contaminated with Pb, Cd, Zn, and Cu. However, the upper soil layer had effectively retained the metals and there had been little vertical movement. In the case of Pb, it was found that 41% was bound by soil organic matter and 28% by carbonate minerals and/or noncrystalline Fe oxides, 8% was relatively mobile on exchange sites, and 4% was residual (nonextractable).

In addition to a strong affinity for Pb, soils are also expected to have large capacities for immobilization of Pb. In a laboratory test of 18 soils, Zimdahl and Skogerboe[4] measured immobilization capacities for Pb between 1.5 to 35×10^{-5} mol Pb/g soil (3 to 70 mg Pb/g soil). Fixation of Pb in these soils was primarily attributed to adsorption onto solid organic matter with precipitation of carbonate minerals and adsorption onto hydrous metal oxides of secondary importance. Rai et al.[12] summarize the relative adsorption capacities of individual soil constituents for Pb as decreasing in the order Mn oxides, Fe oxides, organic matter, and clay minerals. At any given site the total capacity attributable to any one constituent will also be a function of the concentration of that constituent in the solid phase.

Gregson and Alloway[5] showed that in soils contaminated with lead by three different processes (natural weathering and dispersion from an ore body, mining and smelting, and sewage sludge amendment to land) a high-molecular-weight organo-lead complex was present in the soil solution. Low-molecular-weight organic peaks were also accompanied by elution of Pb in all samples except that from the low-pH site. It appeared that the association of Pb with organic matter was greatest in samples of soil solutions with higher pH. Organo-lead complexation may enhance the mobility of lead in the soil zone where the highest level of reactive organic matter is present.

Beneath the soil zone, where organic matter is less abundant and less reactive, it is likely that metal oxides of Fe and Mn play the greatest role in the adsorption of Pb. Kinniburgh et al.[32] studied the adsorption of Pb and other metals onto freshly precipitated Fe gels (amorphous iron hydrous oxide). They found the following selectivity sequence where the numbers in parentheses indicate pH $\pm$ 0.2 for 50% retention: Pb (3.1) > Cu (4.4) > Zn (5.4 > Ni (5.6) > Cd (5.8) > Co (6.0) > Sr (7.4) > Mg (7.8). Lead was very strongly adsorbed onto the iron solid at a low pH and was not released as the pH was raised to 10. Only under very acidic conditions is it expected that adsorption onto this type of iron material would not contribute to immobilizing Pb.

Although Pb itself is not redox sensitive in the natural environment, its mobility may be affected by the redox state of the system. Because Pb is adsorbed onto naturally occurring metal (Fe and Mn) oxyhydroxides, the presence and stability of these solids will effect Pb movement. The adsorption of lead onto these solids will retard the movement of Pb compared with the groundwater flow velocity, unless contamination produces reducing conditions in which case ferric hydroxide and/or manganese oxide/oxyhydroxide may dissolve and release adsorbed lead. If the groundwater becomes reoxidized, the Fe/Mn solids will precipitate and scavenge Pb from solution.

If the degree of Pb contamination is great enough and the environmental conditions are suitable, Pb minerals may form in the subsurface. In a study of lead addition to soil, Santillan-Medrano and Jurinak[6] identified potential solid phases that might limit Pb solution concentrations. In noncalcareous soils, the dissolved Pb concentration appeared to be regulated by $Pb(OH)_2$, $Pb_3(PO_4)_2$, $Pb_4O(PO_4)_2$, and $Pb_5(PO_4)_3OH$ depending on the pH. In groundwater situations, the phosphate concentration is generally low (<1 mg/L), and the formation of phosphate minerals may not provide an effective limit on dissolved Pb. In calcareous soils Santillan-Medrano and Jurinak found that $PbCO_3$ might limit Pb in solution.

Sulfur minerals of Pb may also provide limits on solution concentration. Under oxidizing conditions at high sulfate levels, $PbSO_4$ may form, and under very reducing conditions with sulfide present, the insoluble mineral galena (PbS) may precipitate, thereby limiting the dissolved Pb concentration. The formation of PbS may have occurred at a historical tannery near Woburn, Massachusetts.[33]

In summary, lead is one of the least mobile of the common metal contaminants in the environment. In a study by Korte et al.[34] of the mobility of 11 trace elements (As, Be, Cd, Cr, Cu, Hg, Ni, Pb, Se, V, and Zn) applied to 10 separate soil types, Pb was the second least mobile metal, exceeding only the mobility of copper. Lead is adsorbed onto a wide variety of soil and aquifer solids (organic and inorganic) that appear to have a strong affinity and large capacity for this metal. Several reactive minerals may also limit the solution concentration of lead and affect its mobility. Under alkaline conditions these include $Pb(OH)_2$, $PbSO_4$, and $PbCO_3$. Under reducing conditions PbS may form. As with most metals, if the contaminant source is very acidic (pH < 3) and the environment does not have the capacity to neutralize the acid, then Pb can be relatively mobile.

Chromium

The dominant inorganic chromium species in groundwater are shown in Figure 11-1 along with the stability field for the chromium solid $Cr(OH)_3$. In the range of natural groundwater pH (6.5 to 8.5) under reducing to slightly oxidizing redox potential, cationic and neutral species of Cr(III) ($CrOH^{2+}$, $Cr[OH]_2^+$, and $Cr[OH]_3^0$) dominate with the solid $Cr(OH)_3$ stable over a large portion of the pH/Eh region. Under more oxidizing conditions, the Cr(VI) anionic species $HCrO_4^-$ and CrO_4^{2-} are dominant.[12]

James and Bartlett[35] found that Cr(III) complexed with organic acids (citric acid, diethylenetriaminepentaacetic acid [DTPA], fulvic acid) and water-soluble organic matter thereby enhancing the mobility of chromium. These organic complexants are present naturally in soil and may also occur in sewage sludge, animal manures, and industrial wastewater.

Cr(III) exhibits typical cationic sorption behavior. Adsorption increases with pH as the adsorbent surface sites become more negatively charged and attractive to cations. Specific adsorption of Cr(III) onto Fe and Mn oxides likely occurs under oxidizing conditions.[34] Solid organic matter may also be an important sorbent, although complexation with dissolved organic ligands may reduce adsorption. Because of the low solubility of the reactive Cr(III) solid $Cr(OH)_3$, the effect of Cr(III) adsorption/desorption on chromium mobility is not generally considered an important factor.

The dominant Cr(VI) species chromate (CrO_4^{2-}) exhibits typical anionic sorption behavior. Adsorption decreases with increasing pH as the sorbent surfaces become more negatively charged. Adsorption also decreases when competing dissolved anions are present. Adsorption onto iron oxyhydroxide has been shown to be suppressed by 50 to 80% in the presence of typical groundwater solution concentrations of HCO_3^- and SO_4^{2-} compared with dilute solutions.[36] Orthophosphate (PO_4^{3-}) also competes for adsorption sites with chromate.[37] Adsorption of chromate is not affected by the presence of cations. Even at very high adsorption densities for the metals Cd, Cu, Co, and Zn, Benjamin[38] found no competitive effect for the surface sites on $Fe_2O_3 \cdot H_2O(am)$. Hexavalent chromium adsorbed by nonspecific (exchangeable) processes has been found to be readily desorbed by Cr-free water[39]; however, Cr(VI) bound by stronger specific adsorption bonds was not easily desorbed. This may be partially due to the reduction of Cr(VI) and precipitation of $Cr(OH)_3$.

Oxidation-reduction processes play a major role in affecting the mobility of chromium because it is relatively mobile as Cr(VI) and immobile as Cr(III). It has been shown that where abundant reactive organic matter is present such as in the soil zone, Cr(VI) can be reduced to Cr(III).[37] Under somewhat acidic conditions, it has been found that Cr(VI) reduction by Fe(II) is effective. In this case the reduction of chromium may be followed by the precipitation of the low-solubility solid $(Fe,Cr)(OH)_3(am)$, depending on the solution pH.[40] The natural sources of ferrous iron in the environment that may reduce chromate include residual amounts of reduced iron in minerals such as hematite and biotite. If the source of ferrous iron for chromium reduction is minerals present in the subsurface, then the rate of reduction will be controlled by the rate of release of the reduced iron from the mineral structure.[41] Other naturally occurring reductants such as dissolved sulfides have also been shown to be capable of reducing Cr(VI).[42]

The rate of reduction of Cr(VI) to Cr(III) increases with decreasing pH.[25,37,43] At pH values near 5, the half-life of Cr(VI) (time for half of the initial amount to be reduced) is on the order of one month, whereas at a pH of 7 the half-life may be several years. In cases where ferrous iron is the reductant, the pH effect may be due to the greater solubility of iron minerals under low-pH conditions and the resulting higher dissolved concentration of Fe^{2+}.

In general, the redox condition of the subsurface favors the reduction of Cr(VI) to Cr(III); however, it has been shown that oxidized manganese in the form of MnOOH and MnO_2 can oxidize Cr(III) to Cr(VI).[17] This oxidation is fairly slow under slightly acid to basic conditions, presumably limited by the low solubility of $Cr(OH)_3$.[44] Cr(III) does not appear to be significantly oxidized by the presence of dissolved oxygen in natural water.[42,44]

In summary, the movement of Cr(VI) contamination in the subsurface will be initially retarded by adsorption reactions with exchangeable sites on clay minerals and specific adsorption sites on metal oxide and oxyhydroxides. Other anions in solution will compete for these adsorption sites. If the adsorption capacity is consumed by a high degree of contamination, Cr(VI) will be mobile. Under typical subsurface conditions with a slightly reduced Eh (<250 mV), Cr(VI) will be reduced to Cr(III), which will precipitate as the insoluble mineral $Cr(OH)_3$ or $(Fe,Cr)(OH)_3$. The rate of reduction will be dependent on the pH of the environment and the presence of reductants such as organic matter, Fe^{2+}, and sulfide. If the oxidizing capacity of the chromium contamination exceeds the reducing capacity of the subsurface, Cr(VI) will be relatively stable and mobile. Oxidized manganese minerals may also enhance the stability of Cr(VI) in the environment.

Arsenic

The mobility of arsenic in the subsurface is a function of the species of arsenic present in the environment, the presence of adsorbing surfaces, and the solubility of As minerals in the pH/Eh conditions and solution composition of the aquifer system.

Under oxidizing to mildly reducing conditions in the pH range of 4 to 9, the As(V) species $H_2AsO_4^-$ and $HAsO_4^{2-}$ are dominant, whereas under more reducing conditions the As(III) species $H_3AsO_3^0$ is present at highest concentration (Figure 11-2). Metal oxyhydroxides such as ferric hydroxide, which are common weathering products found in aquifers, have a very strong affinity for the As(V) species and, like adsorption of other oxyanions (e.g., PO_4^{3-} and SeO_4^{2-}), this affinity increases with decreasing pH.[45] These solids also have a very high adsorption capacity for As(V). Fuller et al.[46] measured adsorption densities as high as 0.25 mole As per mole of freshly precipitated ferrihydrite. It is expected that the adsorption capacity would decrease as the precipitate ages and becomes more crystalline, reducing the number of surface sites. As(III) species are not as strongly attracted to ferrihydrite as the As(V) species. Unlike As(V), the adsorption of As(III) onto iron hydroxide increases with pH up to a maximum adsorption affinity at a pH of about 7.

The adsorption of arsenic onto ferrihydrite can have a major effect on the cycling of arsenic at a redox interface either in sediments or along the flowpath of an aquifer. Under oxidizing conditions, ferrihydrite will be stable and will accumulate arsenic entering the environment either naturally from weathering or from a contaminant source. Arsenic will be in the form of the strongly adsorbed As(V) species. If the environment becomes reducing because of a lack of oxygen and the presence of organic matter, As(V) may be reduced to As(III), which will mobilize some of the arsenic because As(III) is less strongly held by ferrihydrite. If the redox potential decreases further (<100 mV), then the ferrihydrite may no longer be stable and it will dissolve, releasing any attached arsenic. In sediments at a reservoir in Montana, the arsenic pore water concentration increased from 20 μg/L (predominantly As[V]) in the upper oxidized layer to 550 μg/L (predominantly As[III]) in a lower reduced layer.[47] The dissolved arsenic in sediment pore water may diffuse upward to more oxidizing conditions and re-adsorb onto ferric hydroxide.[48] In an aquifer, if reducing conditions persist downgradient of the redox interface, arsenic will remain mobile as an As(III) species unless sulfide is present in which case an insoluble arsenic sulfide mineral may limit the dissolved arsenic concentration.

The presence of the common groundwater anions chloride, nitrate, and sulfate at concentrations of 10^{-4} to 10^{-2} mol/L have little effect on the adsorption of As. However, phosphate has been shown to substantially suppress As adsorption by soils.[49] The methylation of arsenic to monomethylarsonic acid (MMAA) and dimethylarsinic acid (DMAA) may increase the mobility of arsenic because these species are less strongly adsorbed than inorganic As(V).[50,51] However, MMAA and DMAA are rarely the dominant arsenic species in the environment. Even in the case of contamination by tannery waste at a site in Massachusetts where the groundwater had a dissolved organic carbon concentration as high as 460 mg/L, the proportion of methylated arsenic species to total arsenic was generally less than a few percent with a maximum value of 26%.[33]

Very few arsenic minerals appear to form in the soil or aquifer environment and provide a potential solubility control on dissolved arsenic concentration. Under oxidizing conditions, the mineral scorodite ($FeAsO_4 \cdot 2H_2O$) has been reported by several investigators[24,29,51]; however, its stability field appears to be limited to fairly acidic conditions (between pH 4 and 5). At the Ashanti Gold mine in Ghana, Bowell et al.[51] report the occurrence of bukovskyite ($Fe_2AsO_4OH \cdot 7H_2O$), kankite ($FeAsO_4 \cdot 3.5H_2O$), and pittcite (amorphous hydrated ferric arsenate) as secondary arsenic minerals in addition to scorodite. Under reducing conditions, it appears likely that orpiment (As_2S_3) and/or realgar (AsS) may form if sulfide is present.

In summary, the mobility of arsenic under oxidizing conditions is primarily affected by the adsorption of As(V) onto metal oxyhydroxide surfaces. If the appreciable adsorption capacity of these surfaces is not surpassed, then arsenic movement will be strongly retarded because of the high affinity of these surfaces for As(V). Under reducing conditions, the dominant arsenic redox species will be As(III), which is not as strongly adsorbed. Furthermore, the primary adsorbing solids may not be stable if the redox potential is low enough. As a consequence, arsenic is expected to be much more mobile under reducing conditions. This mobility may be significantly reduced if arsenic sulfide minerals become saturated and precipitate.

11.5. REMEDIATION

Removal or immobilization of Pb, Cr, or As in an aquifer for restoration purposes will be affected by the geochemical reactivity of the metal in the contaminated environment. At some sites the insolubility of solids containing the contaminant may make immobilization a viable restoration method, whereas under other site conditions the mobility of the metal and its equilibrium solution concentration will require removal of at least part of the mass of the subsurface contaminant.

Lead

The common immobility of lead in the environment inhibits its removal, but enhances the application of *in situ* fixation techniques. If the solution concentration of Pb is limited by specific adsorption or the presence of insoluble minerals, then pump-and-treat removal is unlikely to be successful unless the degree of restoration required is very small (i.e., a small amount of lead must be removed to achieve the cleanup goal). If a large proportion of the adsorbed Pb is present as an exchangeable species or if the lead minerals are relatively soluble, then removal of Pb from the subsurface is a possibility. The removal of exchangeable Pb during a pump-and-treat operation can be enhanced by the addition of other divalent cations (e.g., Ca^{2+}, Mg^{2+}) in relatively high concentration to the water circulated through the zone of contamination. These cations will replace Pb on the surface sites allowing it to be more easily flushed from the system.

If a relatively soluble lead solid is providing a source of contamination to the groundwater, then its solubility can be increased to more rapidly flush it from the system. For example, Paulson et al.[52] found that $PbSO_4$ was present in the subsurface at a former battery recycling facility. When the solid is a sulfate mineral, a complexing agent such as Mg^{2+} can be added to the water to increase the solubility of the mineral because of the formation of $MgSO_4^0$ in solution, which lowers the effective concentration of SO_4^{2-}. Alternatively, barium could be added to the water to precipitate sulfate as barite ($BaSO_4$) also increasing the solubility of $PbSO_4$. If the solid is a carbonate mineral, then its solubility is highly pH dependent and manipulating the pH can effectively increase solubility and removal rate. The solubility of sulfate minerals is not directly pH dependent. Knowing the identity of the solid controlling solution concentration of the contaminant aids in the development of an efficient remediation technique.

If removal of lead from the aquifer does not appear to be an efficient restoration method, then *in situ* fixation may be considered. The low solubility of PbS under reducing conditions may provide for aquifer restoration if the natural system is reducing with respect to sulfur. The addition of sulfide (as Na_2S or H_2S) to the zone of contamination should precipitate PbS. The kinetics of precipitation under simulated site conditions should be evaluated in the laboratory to ensure that the precipitation rate can remove Pb in an acceptable time frame. As with any immobilization restoration technique, the potential dissolution of PbS under natural conditions that will be reestablished after active restoration has ceased must be evaluated.

The use of the funnel-and-gate system for treating groundwater contaminated with lead has also been suggested. One approach consists of adding a relatively soluble mineral such as hydroxyapatite ($Ca_3[PO_4]_2$) to the reactive gate material. The hydroxyapatite will dissolve into the Pb contaminated groundwater flowing through the gate and precipitate and immobilize lead as a phosphate mineral.[53] Several reactive lead phosphate minerals ($Pb_3[PO_4]_2$, $Pb_4O[PO_4]_2$, and $Pb_5[PO_4]_3OH$) that might form in the reaction wall have been identified in soil.[6]

Chromium

Cr(VI) is relatively mobile in most environments because its solids are soluble and it is not strongly adsorbed; therefore it is a potential candidate for removal from an aquifer by pump-and-treat. However, the adsorption affinity of the solid phase for Cr(VI) increases with decreasing chromium concentration, making it more difficult to remove as the concentration decreases during pump-and-treat remediation.[26] Extraction of chromate from an aquifer could be enhanced by the addition of other anions such as sulfate, bicarbonate, or phosphate, which compete for the adsorption sites and would mobilize CrO_4^{2-}. Complicating the removal of Cr(VI) is the fact that it is not the stable redox state of chromium in most aquifer pH/Eh conditions, and the natural reduction of Cr(VI) to Cr(III) and precipitation of $Cr(OH)_3$ tends to immobilize chromium. For this reason, pump-and-treat efforts to remove total chromium from the subsurface are not expected to be efficient. The addition of a strong oxidizing agent such as hydrogen peroxide to an aquifer has been shown to increase the removal rate of chromium; however, the enhancement was not considered significant compared with the total amount of chromium requiring removal.[54]

The solid-phase form of chromium contamination in the environment should be considered in designing a remediation scheme. At sites heavily contaminated with Cr(VI) it has been shown that chromium may be present as the solid $BaCrO_4$. To remediate a site with a relatively soluble chromate mineral such as barium chromate, the solid must be flushed from the system before significant decreases in solution concentration of the contaminant will be possible (the "tailing effect," Section 8.3). Column study results have shown that the addition of sulfate to water flowing through the zone of $BaCrO_4$ contamination can increase the solubility of the mineral by an order of magnitude, thereby reducing the amount of water required for flushing the system.[55] The addition of sulfate lowers the solution concentration of barium by precipitating the mineral barite ($BaSO_4$). The depressed solution concentration of barium increases the solubility of $BaCrO_4$. Released chromate would need to be flushed from the system before the chromium is reduced to Cr(III) and precipitated as $Cr(OH)_3$.

From a geochemistry standpoint, an efficient method for removing chromium from contaminated groundwater is *in situ* precipitation and immobilization in the aquifer as $Cr(OH)_3$. The solubility of this solid in the pH range of 6 to 9 is less than the chromium drinking water standard (50 µg/L). The reduction of Cr(VI) to Cr(III) and precipitation of this mineral is relatively slow at pH values greater than about 5; therefore various method of catalyzing the reduction have been proposed and tested. Laboratory results show that the reduction of Cr(VI) by ferrous iron occurs rapidly even in the presence of dissolved oxygen.[56] James[57] found in batch experiment that Mn^{2+}, Fe^{2+}, steel wool (Fe^0), leaf litter, and lactic acid reduced various quantities of Cr(VI) in soils.

Grove and Ellis[58] suggested that Cr(VI) spills could be treated by acidification and addition of a reducing agent such as leaf litter. Lowering the pH increases the reduction reaction rate and the organic matter enhances the reducing environment. After reduction of Cr(VI) has been achieved, the site would need to be neutralized to lower the solubility of $Cr(OH)_3$. A potential unwanted side effect of this process might be increased, dissolved organic carbon and complexation with metals facilitating their transport. Shallow Cr(VI) contamination has been successfully immobilized at a site on the Delaware River where a ferrous sulfate heptahydrate solution was injected into the aquifer to reduce Cr(VI) to Cr(III).[59] Groundwater concentrations of about 85 mg/L Cr(VI) were successfully reduced to below 50 µg/L across most of the site during an application time of less than 1 month. Residual zones of elevated chromium concentration exist where application of the treatment chemical was inhibited by structures on the site and the fluctuating water table. Reductants in addition to ferrous sulfate that have been shown in the laboratory to successfully reduce chromium include sodium sulfide and hydrogen sulfide.[60]

For contamination of Cr(VI) in the vadose zone, which may leach into an aquifer, reducing gases such as sulfur dioxide and hydrogen sulfide have been suggested to convert the mobile Cr(VI) to Cr(III).[60] Laboratory tests have shown that 90% of the Cr(VI) at an initial soil concentration of several hundred mg/L can be immobilized by treatment with 100 ppm hydrogen sulfide gas for a period of one day.[61]

In the design of a remediation scheme to reduce and immobilize chromium *in situ*, the oxidation capacity of the system must be considered because it will consume reductants added to the subsurface. It has been shown that the major oxidant for Cr(III) in the environment is oxidized Mn minerals. Bartlett and James[17] have developed a test for determining the oxidation ability of the subsurface material to convert Cr(III) to Cr(VI). The relative ability of material to oxidize Cr can also be predicted by measuring the amount of Mn(IV) present in subsurface solids with hydroquinone. Dissolved oxygen and nitrate may also consume some of the ferrous iron or sulfide added in aqueous form to reduce Cr(VI).

Dissolved Cr(VI) in groundwater may also be treated by interception at a permeable reaction gate. Several investigators[62-64] have found in the laboratory that zero-valent iron (Fe^0) is an effective reductant for chromate. This approach, known as the Waterloo process, is being tested at a chromium contaminated location associated with the U.S. Coast Guard Station near Elizabeth City, New Jersey.

Arsenic

The removal of arsenic from an oxidized soil or aquifer is enhanced by the high solubility of likely arsenate minerals, but is inhibited by the strong adsorption of As(V) onto metal oxyhydroxide minerals that are present in most oxidized environments. As mentioned above in the discussion of contaminant migration, scorodite ($FeAsO_4 \cdot 2H_2O$) appears to be the only common arsenate mineral that might form under oxidizing contaminant conditions. However, this mineral is relatively soluble except in a narrow pH range of 4 to 5. Under more alkaline conditions typical of most soils and aquifers, it should be relatively easy to dissolve this mineral, if present, and mobilize arsenic. The very strong affinity and high capacity of ferrihydrite and other metal oxyhydroxides for As(V) will make flushing arsenic from the system a challenge unless the adsorption capacity is exceeded.

On the one hand, strong As(V) adsorption hinders removal of arsenic, but conversely adsorption can severely limit arsenic solution concentration. A balance must be reached between acceptable concentrations in the soil solid phase and the shallow groundwater. If the adsorption capacity does not exceed the soil cleanup level and the equilibrium solution concentration does not exceed the groundwater cleanup level for a site, then the immobility of As(V) is an advantage. The major problem comes if either cleanup level is exceeded. In this occurs, simple pump-and-treat will require many pore volumes of fresh water to be pumped through the zone of contamination because very little As(V) will be desorbed during each cycle. The properties of arsenic can be used to enhance pump-and-treat restoration for the oxidized zone. Although most anions do not successfully compete for the As(V) adsorption sites, phosphate is a strong competitor,[11] and if sulfate is added to the water at a high enough concentration, it will also remove some As(V).

Treating the adsorbent directly may also enhance restoration. The metal oxides may be dissolved to release arsenic by lowering the pH or Eh of the system to a level at which the solid is soluble.[65] This may be difficult to achieve if the system has a high buffering capacity. If only the groundwater cleanup level is exceeded, adsorption might be enhanced by adding a solution with high dissolved iron and precipitating additional ferrihydrite. This should scavenge more arsenic from solution, reducing the solution concentration to the cleanup level.

Under reducing conditions, arsenic will be relatively mobile unless sulfide is present. If arsenic sulfide minerals are limiting the arsenic concentration in solution, then it is likely that the concentration will be below the cleanup level because of the low solubility of these minerals. If sulfide minerals are not present, then arsenic will be relatively mobile and may be amenable to pump-and-treat. To immobilize arsenic in a sulfide-poor reducing environment, sulfide could be added as Na_2S or H_2S. This should enhance the precipitation of the arsenic sulfide minerals. Alternatively, the reduced system could be oxidized, and any initially present or added iron in solution would precipitate as ferrihydrite and remove arsenic from solution by adsorption. In this case it would be necessary to ensure that the established oxidizing condition is permanent and that the system does not revert to a reducing environment after the cessation of remediation.

REFERENCES

1. Davis, B.E., in *Heavy Metals in Soils*, Alloway, B.J., Eds., Blackie, Glascow and London, 1990, pp. 177–196.
2. Miller, W.P. and McFee, W.W., Distribution of cadmium, zinc, copper and lead in soils of industrial northwestern Indiana, *J. Environ. Qual.*, 12, 29–33, 1983.
3. Lindsay, W.L., *Chemical Equilibria in Soils,* John Wiley, New York, 1979.
4. Zimdahl, R.L. and Skogerboe, R.K., Behavior of lead in soil, *Environ. Sci. Technol.*, 11, 1202–1207, 1977.
5. Gregson, S.K. and Alloway, B.J., Gel permeation chromatography studies on the speciation of lead in solutions of heavily polluted soils, *J. Soil Sci.*, 35, 55–61, 1984.

6. Santillan-Medrano, J. and Jurinak, J.J., The chemistry of lead and cadmium in soil: solid phase formation, *Soil Sci. Soc. Amer. Proc.*, 39, 851–856, 1975.

7. Borden, R.C. and Yanoschak, T.M., Ground and surface water quality impacts of North Carolina sanitary landfills, *Water Res. Bull.*, 26, 269–277, 1990.

8. McGrath, S.P. and Smith, S., in *Heavy Metals in Soil*, Alloway, B.J., Eds., Blackie, Glasgow and London, 1990, pp. 125–150.

9. Puls, R.W., Clark, D.A., Carlson, C., and Vardy, J., Characterization of chromium-contaminated soils using field-portable x-ray fluorescence, *GWMR*, 111–115, 1994.

10. James, B.R., The challenge of remediating chromium-contaminated soil, *Environ. Sci. Technol.*, 30, 248A–251A, 1996.

11. O'Neill, P., in *Heavy Metals in Soils*, Alloway, B.J., Eds., Blackie, Glasgow and London, 1990, pp. 83–99.

12. Rai, D. and Zachara, J.M., *Chemical Attenuation Rates, Coefficients, and Constants in Leachate Migration: A Critical Review*, Vol. 1, Electric Power Research Institute, Palo Alto, CA, 1984.

13. Thomas, D.J., Arsenic toxicity in humans: research problems and prospects, *Env. Geochem. and Health*, 16, 107–111, 1994.

14. Puls, R.W. and Powell, R.M., Acquisition of representative ground water quality samples for metals, *GWMR*, 12, 167–176, 1992.

15. Aggett, J. and Kriegman, M.R., Preservation of arsenic(III) and arsenic(V) in samples of sediment interstitial water, *Analyst.* 112, 153–157, 1987.

16. Cherry, J.A., Shaikh, A.U., Tallman, D.E., and Nicholson, R.V., Arsenic species as an indicator of redox conditions in groundwater, *J. Hydrol.*, 43, 373–392, 1979.

17. Bartlett, R. and James, B., Behavior of chromium in soils. III. Oxidation, *J. Environ. Qual.*, 8, 31–35, 1979.

18. Allison, J.D., Brown, D.S., and Novo-Gradac, K.J., *MINTEQA2/PROEDFA2, a Geochemical Assessment Model for Environmental Systems*, U.S. Environmental Protection Agency, Athens, GA, 1991.

19. Parkhurst, D.L., *User's Guide to PHREEQC, a Computer Model for Speciation, Reaction Path, Advective Transport and Inverse Geochemical Calculations*, U.S. Geological Survey, Denver, CO, 1995.

20. Loux, N.T., Brown, D.S., Chafin, C.R., Allison, J.D., and Hassan, S.M., Chemical speciation and competitive cationic partitioning on a sandy aquifer material, *Chemical Speciation and Bioavailability*, 1, 111–125, 1989.

21. Dzombak, D.A., *Toward a Uniform Model for the Sorption of Inorganic Ions on Hydrous Oxides*, Massachusetts Institute of Technology, Boston, 1986.

22. Baltpurvins, K.A., Burns, R.C., Lawrance, G.A., and Stuart, A.D., Use of the solubility domain approach for the modeling of the hydroxide precipitation of heavy metals from wastewater, *Environ. Sci. Technol.*, 30, 1493–1499, 1996.

23. Blowes, D.W. and Jambor, J.L., The pore-water geochemistry and the mineralogy of the vadose zone of sulfide tailings, Waite Amulet, Quebec, Canada, *Appl. Geochem.*, 5, 327–346, 1990.

24. Doyle, T.A., Davis, A., and Runnells, D.D., Predicting the environmental stability of treated copper smelter flue dust, *Appl. Geochem.*, 9, 337–350, 1994.

25. Rai, D. et al., *Chromium Reactions in Geologic Material*, Electric Power Research Institute, Palo Alto, CA, 1988.

26. Palmer, C.D. and Wittbrodt, P.R., Processes affecting the remediation of chromium-contaminated sites, *Env. Health Persp.*, 92, 25–40, 1991.

27. Rai, D., Sass, B.M., and Moore, D.A., Chromium(III) hydrolysis constants and solubility of chromium(III) hydroxide, *Inorg. Chem.*, 26, 345–349, 1987.

28. Sass, B.M. and Rai, D., Solubility of amorphous chromium(III)-iron(III) hydroxide solid solutions, *Inorg. Chem.*, 26, 2228–2232, 1987.

29. Davis, A. and Ashenberg, D., The aqueous geochemistry of the Berkeley Pit, Butte, Montana, U.S.A., *Appl. Geochem.*, 4, 23–36, 1989.

30. Hess, R.E. and Blancher, R.W., Dissolution of arsenic from waterlogged and aerated soils, *Soil Sci. Soc. Am. J.*, 41, 861–865, 1977.

31. Ferguson, J.F. and Gavis, J., A review of the arsenic cycle in natural waters, *Water Res.*, 6, 1259–1274, 1972.

32. Kinniburgh, D.G., Jackson, M.L., and Syers, J.K., Adsorption of alkaline earth, transition and heavy metal cations by hydrous oxide gels of iron and aluminum, *Soil Sci. Soc. Am. J.,* 40, 796–799, 1976.

33. Davis, A., Kempton, J.H., Nicholson, A., and Yare, B., Groundwater transport of arsenic and chromium at a historical tannery, Woburn, Massachusetts, U.S.A., *Appl. Geochem.,* 9, 569–582, 1994.

34. Korte, N.E., Skopp, J., Fuller, W.H., Niebla, E.E., and Alesii, B.A., Trace element movement in soils: influence of soil physical and chemical properties, *Soil Science,* 122, 350–359, 1976.

35. James, B.R. and Bartlett, R.J., Behavior of chromium in soils. V. Fate of organically complexed Cr(III) added to soil, *J. Environ. Qual.,* 12, 169–172, 1983.

36. Zachara, J.M., Girvin, D.C., Schmidt, R.L., and Resch, C.T., Chromate adsorption on amorphous iron oxyhydroxide in the presence of major groundwater ions, *Environ. Sci. Technol.,* 2, 589–594, 1987.

37. Bartlett, R.J. and Kimble, J.M., Behavior of chromium in soils. II. Hexavalent forms, *J. Environ. Qual.,* 5, 383–386, 1976.

38. Benjamin, M.M., Adsorption and surface precipitation of metals on amorphous iron oxyhydroxide, *Environ. Sci. Technol.,* 17, 686–692, 1983.

39. Stollenwerk, K.G. and Grove, D.B., Adsorption and desorption of hexavalent chromium in an alluvial aquifer near telluride, Colorado, *J. Environ. Qual.,* 14, 150–155, 1985.

40. Eary, L.E. and Rai, D., Chromate reduction by subsurface soils under acidic conditions, *Soil Sci. Soc. Am. J.,* 55, 676–683, 1991.

41. Eary, L.E. and Rai, D., Kinetics of chromate reduction by ferrous ions derived from hematite and biotite at 25°C, *Am. J. Sci.,* 289, 180–213, 1989.

42. Schroeder, D.C. and Lee, G.F., Potential transformations of chromium in natural waters, *Water, Air and Soil Pollut.,* 4, 355–365, 1975.

43. Henderson, T., Geochemical reduction of hexavalent chromium in the Trinity Sand Aquifer, *Ground Water,* 32, 477–486, 1994.

44. Eary, L.E. and Rai, D., Kinetics of chromium, (III) oxidation to chromium(VI) by reaction with manganese dioxide, *Environ. Sci. Technol.,* 21, 1187–1193, 1987.

45. Pierce, M.L. and Moore, C.B., Adsorption of arsenite and arsenate on amorphous iron hydroxide, *Water Res.,* 16, 1247–1253, 1982.

46. Fuller, C.C., Davis, J.A., and Waychunas, G.A., Surface chemistry of ferrihydrite. II. Kinetics of arsenate adsorption and coprecipitation, *Geochim. Cosmochim. Acta,* 57, 2271–2282, 1993.

47. Moore, J.N., Ficklin, W.H., and Johns, C., Partitioning of arsenic and metals in reducing sulfidic sediments, *Environ. Sci. Technol.,* 22, 432–437, 1988.

48. Belzile, N. and Tessier, A., Interactions between arsenic and iron oxyhydroxides in lacustrine sediments, *Geochim. Cosmochim. Acta,* 54, 103–109, 1990.

49. Livesey, N.T. and Huang, P.M., Adsorption of arsenate by soils and its relation to selected chemical properties and anions, *Soil Science,* 131, 88–94, 1981.

50. Holm, T.R., Anderson, M.A., Iverson, D.G., and Stanforth, R.S., in *Chemical Modeling in Aqueous Systems,* Jenne, E.A., Eds., American Chemical Society, Washington, DC, 1979, pp. 711–736.

51. Bowell, R.J., Sorption of arsenic by iron oxides and oxyhydroxides in soils, *Appl. Geochem.,* 9, 279–286, 1994.

52. Paulson, S.E., Petrie, L.M., and Marozas, D.C., *Geologic-Geochemical Characterization of Heavy Metal Contamination in Soils,* Hazardous Materials Control Resources Institute, Vienna, VA, 1992.

53. Starr, R.C. and Cherry, J.A., In situ remediation of contaminated ground water: the funnel-and-gate system, *Ground Water,* 32, 465–476, 1994.

54. Davis, A. and Olsen, R.L., The geochemistry of chromium migration and remediation in the subsurface, *Ground Water,* 33, 759–768, 1995.

55. Ward, C. and Palmer, C.D., *Sulfate Extraction of Chromate Contaminated Soils: Column Studies of Enhanced Barium Chromate Removal,* National Ground Water Association, Minneapolis, MN, 1994.

56. Eary, L.E. and Rai, D., Chromate removal from aqueous wastes by reduction with ferrous ion, *Environ. Sci. Technol.,* 22, 972–977, 1988.

57. James, B.R., Hexavalent chromium solubility and reduction in alkaline soils enriched with chromite ore processing residue, *J. Environ. Qual.,* 23, 227–233, 1994.

58. Grove, J.H. and Ellis, B.H., Extractable chromium as related to pH and applied chromium, *Soil Sci. Soc. Am. J.,* 44, 238–242, 1980.

59. Brown, R.A., Crosbie, J.R., Ramage, T., and O'Neil, S., *In Situ Treatment of Chromium VI with an Iron Reduction Process*, Water Environment Federation 66th Annual Conference, Oct. 3–7, Anaheim, CA, 1993.

60. Thornton, E.C., Jurgensmeier, C.A., and Baechler, M.A., *Laboratory Evaluation of the In Situ Chemical Treatment Approach for Remediation of Cr(VI)-Contaminated Soils and Groundwater,* Westinghouse Hanford Company, Richland, WA, 1991.

61. Thornton, E.C. and Trader, D.E., *Laboratory Evaluation of the In Situ Chemical Treatment Approach to Soil and Groundwater, Remediation,* Proceedings Environmental Restoration Conference, Augusta, GA, Oct.24–28, 1993, sponsored by U.S. Depaartment of Energy, Office of Environmental Restoration, Washington, DC.

62. Gillham, R.W. and Burris, D.R., *Recent Developments in Permeable In Situ Treatment Walls for Remediation of Contaminated Groundwater,* Proceedings from Subsurface Restoration Conference, 3rd International Conference on Groundwater Quality Research, June 21–24, Dallas, TX, 1992.

63. Blowes, D.W. and Ptacek, C.J., *Geochemical Remediation of Groundwater, by Permeable Reactive Walls: Removal of Chromate by Reaction with Iron-Bearing Solids,* Dallas, TX, 1992.

64. Powell, R.M., Puls, R.W., and Paul, C.J., *Chromate Reduction and Remediation Utilizing the Thermodynamic Instability of Zero-Valence State Iron,* Water Environment Federation, Miami, FL, 1994.

65. Xu, H., Allard, B., and Grimvall, A., Effects of acidification and natural organic materials on the mobility of arsenic in the environment, *Water, Air, Soil Pollut.,* 57–58, 269–278, 1991.

12 GEOCHEMISTRY OF ACID MINE WASTE

One of the more common types of environments in which inorganic contaminants are present at levels of concern is a site with acidic (low-pH) conditions. As will be discussed in this chapter, minerals are relatively soluble under acidic conditions and adsorption is not as effective a mechanism for removing contaminants from solution. For these reasons contaminants are relatively mobile at high concentrations posing a potential threat to human health and the environment. The common processes that generate acid in the environment are discussed in this chapter, as well as water/rock interactions that can neutralize acid. The environments where acidic conditions occur such as mine sites and tailings dumps are described. Sampling considerations, geochemical modeling, contaminant transport, and remediation of acid-impacted environments are also discussed.

12.1. ACID-GENERATING AND -NEUTRALIZING PROCESSES

Acidic conditions can occur naturally or be produced directly or indirectly by human activities. The disposal of a strong acid (e.g., sulfuric [H_2SO_4], nitric [HNO_3] or hydrochloric [HCl]) is the most direct, and least common, method of creating a localized acidic environment. More common conditions are the natural or enhanced oxidation of pyrite and other sulfide minerals in ore or coal deposits, storage piles, or minespoils. Enhanced oxidation can be caused by mining, storage, milling, and tailings disposal that facilitates exposure of sulfide minerals to oxidizing earth surface conditions. Widespread acidic conditions can result from the production of acidic gases (primarily SO_x and NO_x) from coal combustion and auto exhaust. The acid precipitation that can form in the atmosphere may increase the acidity of the downwind environment. This discussion of acid generation will focus on the acidic environments created by natural or man-induced oxidation of sulfide minerals as representative of acid-type contaminant conditions. The pH neutralization reactions for all types of acid input will be similar to that described for the sulfuric acid case.

Acid-Producing Reactions

Pyrite (FeS_2) and other sulfide minerals form under reducing conditions where sulfide (S[-II]) is the dominant redox form of sulfur. Sulfide minerals are not stable in the presence of molecular oxygen (O_2); therefore they will oxidize and dissolve if exposed to earth surface conditions or groundwater with dissolved oxygen (DO). Figure 12-1 is a pH-Eh diagram of the $Fe-S-H_2O$ system that shows the stability fields of pyrite and the Fe(III) minerals ferrihydrite ($Fe[OH]_3$) and jarosite ($KFe_3[SO_4]_2[OH]_6$). Under reducing conditions pyrite is stable, but as the system becomes more oxidizing ferrihydrite replaces pyrite as the stable iron mineral. Ferrihydrite is stable over a very wide range of pH and Eh conditions, which is the reason it is such a common weathering product of primary iron-bearing minerals. Under acidic, highly oxidizing conditions, with sufficient sulfate present, jarosite becomes the stable iron mineral. The fields for Fe^{2+} and Fe^{3+} on Figure 12-1 represent regions where the iron minerals are relatively soluble at dissolved concentrations greater than 10^{-6} mol/L.

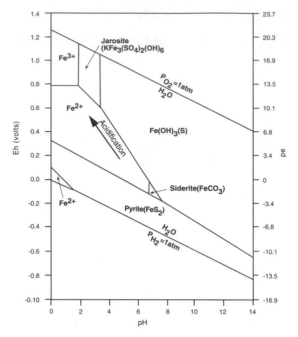

Figure 12-1 Stability fields of iron minerals ($a_{Fe^{2+}}$ = a_{k^+} = 10^{-4} M; a_s = 10^{-2} M).

In an aquifer containing pyrite, the reducing capacity of the sulfide minerals may be sufficient to create anoxic conditions and low DO levels and to produce a low redox potential if the flux of oxidizing groundwater is not high. However, in the unsaturated zone and on the earth surface, the reservoir of oxygen in the air can maintain oxidizing conditions and continue oxidation of the sulfide minerals.

The oxidation of pyrite to its stable products under oxidizing conditions is not a simple process. A general overall reaction can be written as follows:

$$FeS_2 + 15/4\,O_2 + 7/2\,H_2O \rightarrow Fe(OH)_3 + 2SO_4^{2-} + 4H^+ \qquad (12\text{-}1)$$

This reaction shows that in the presence of molecular oxygen Fe(II) and S_2(-II) in pyrite are oxidized by O_2 to produce the stable Fe(III) mineral ferrihydrite (Fe[OH]$_3$) and dissolved sulfate (S[VI]) and hydrogen ions. The release of hydrogen with a strong acid anion (sulfate) results in an acidic solution unless other reactions occur to neutralize the hydrogen ions. Singer and Stumm[1] showed that oxygen is a poor direct oxidant for pyrite and that Reaction 12-1 is propagated at low pH by the following two coupled reactions:

$$FeS_2 + 14Fe^{3+} + 8H_2O \rightarrow 15Fe^{2+} + 2SO_4^{2-} + 16H^+ \qquad (12\text{-}2)$$

$$14Fe^{2+} + 7/2\,H_2O + 14H^+ \rightarrow 14Fe^{3+} + 7H_2O \qquad (12\text{-}3)$$

Presenting the reactions in this form illustrates the importance of ferric iron as the primary oxidant of sulfide, with oxygen as the oxidant for ferrous iron. Working in unison, ferric iron oxidizes the sulfide in pyrite, while oxygen replaces the reduced iron by oxidizing ferrous iron to ferric. Moses et al.[2] also provide evidence that supports the primary role of ferric iron as the direct oxidant of pyrite in aerobic and anaerobic systems. Moses and Herman[3] extended the Singer-Stumm model for pyrite oxidation to include the neutral pH region. In this pH region they propose that ferrous iron adsorbs onto the pyrite surface and blocks direct oxidation by ferric iron. The adsorbed ferrous iron is oxidized to ferric iron to allow the oxidation of pyrite to continue.

The concentration of the oxidant ferric iron in solution is limited by the formation of the mineral ferrihydrite:

$$Fe^{3+} + 3H_2O = Fe(OH)_3(s) + 3H^+ \qquad (12\text{-}4)$$

The precipitation of ferrihydrite not only places a limit on available dissolved ferric iron, but it also provides significant acidity to the solution by the release of hydrogen ions from water as the solid forms.

The rate-limiting step for the oxidation of pyrite has been shown to be the oxidation of ferrous iron to ferric iron (Reaction 12-3). Singer and Stumm[1] demonstrated that the direct oxidation of pyrite by air was too slow to produce the amount of acidity observed by oxidation of pyrite in the environment, and they suggested that the iron-oxidizing bacterium Thiobacillus *ferrooxidans* acts as a catalyst in nature to facilitate the oxidation of iron to the ferric state to drive the pyrite oxidation reaction. Kleinmann et al.[4] showed that, as the pH falls in the region of 5.5 to 4.5, the rate of abiotic oxidation of ferrous iron decreases and slows pyrite oxidation. At this point *T. ferrooxidans* takes on a primary role of oxidizing ferrous iron and catalyzing oxidation of pyrite. The decreasing pH, increasing Eh, and microbial activity all work to increase Fe^{3+} in solution, enhancing the speed of pyrite oxidation. The chemical evolution of the acidic water in shown on Figure 12-1. At the point in which ferrihydrite is relatively soluble, the limiting factor in the rate of pyrite oxidation is the availability of dissolved oxygen to oxidize ferrous iron released from the pyrite. On the earth surface or the unsaturated zone, this is not a serious limitation; however, below the water table the supply of molecular oxygen will be limited by its relatively low solubility (about 8 mg/L) in water. Kleinmann and Crerar[5] demonstrated that *T. ferrooxidans* is only effective at catalyzing the oxidation of pyrite in the unsaturated zone and not in saturated environments.

Acid-Neutralizing Reactions

The oxidation of pyrite (Reaction 12-2) and the precipitation of ferrihydrite (Reaction 12-4) release hydrogen to solution and increase the solution's acidity unless the hydrogen is neutralized by other water/rock interactions. The major reservoir of neutralizing capacity in the environment is the silicate minerals, which make up the majority of the minerals in the earth's crust. Chemical weathering of these minerals consumes dissolved hydrogen and can be represented as follows:

$$MeAlSiO_n(s) + 4H^+ \rightarrow Me^+ + Al^{3+} + H_4SiO_4^0 \qquad (12\text{-}5)$$

where Me is Ca, Na, K, Mg, or Fe.

The primary minerals may be pure silicates such as fayalite (Fe_2SiO_4) or enstatite ($MgSiO_3$) or aluminosilicates such as albite ($NaAlSi_3O_8$) or biotite ($KFe_3AlSi_3O_{10}[OH]_2$). Weathering of albite under acid conditions can be represented as follows:

$$NaAlSi_3O_8 + H^+ + 7H_2O \rightarrow Na^+ + Al(OH)_3(s) + 3H_4SiO_4^0 \qquad (12\text{-}6)$$

In this reaction each mole of albite dissolved consumes one mole of hydrogen ions in solution. The byproducts of albite weathering are Na^+, silicic acid ($H_4SiO_4^0$), and the mineral gibbsite ($Al[OH]_3$). Silicic acid is a very weak acid and does not contribute significant H^+ to solution unless the pH is greater than 9. The presence of gibbsite as a solid phase provides additional neutralizing ability to the system because it can consume dissolved hydrogen by the following reaction:

$$Al(OH)_3(s) + 3H^+ \rightarrow Al^{3+} + 3H_2O \tag{12-7}$$

The chemical weathering rate of the primary silicate and aluminosilicate minerals is slow relative to the production rate of acid by pyrite oxidation under earth surface conditions and is slow even under conditions of typical groundwater residence times of tens to hundreds of years. In general, aluminum substitution into the silicate framework makes the mineral more susceptible to acid attack, and geologic evidence shows that anorthite ($CaAl_2Si_2O_8$) is more sensitive to chemical weathering than albite ($NaAlSi_3O_8$) and that quartz (SiO_2) is the least sensitive silicate mineral to acid weathering.[6] Compared with the weathering rates of even the most reactive primary silicate minerals, the reaction rates of secondary minerals (such as gibbsite and clays) and sedimentary minerals (such as carbonates) are relatively rapid and can neutralize acid produced by or introduced into the environment.

Because of its common occurrence in a wide variety of environments and rapid rate of reaction, calcite is the most important neutralizing agent in the subsurface. Calcite may occur as a weathering product of almost any rock type or an original mineral in a sedimentary rock (e.g., limestone). The formation of calcite from the weathering of a primary aluminosilicate mineral such as anorthite can be shown as follows:

$$CaAl_2Si_2O_8(\text{anorthite}) + CO_2(aq) + 2H_2O = CaCO_3(\text{calcite}) + Al_2Si_2O_5(OH)_4(\text{kaolinite}) \tag{12-8}$$

In this reaction the calcium released by the dissolution of anorthite combines with dissolved carbonate to form calcite ($CaCO_3$). In the unsaturated zone or at the water table, carbonate can be replenished to the water by dissolving additional CO_2(gas) from the soil vapor. The remaining anorthite constituents are shown as forming the clay mineral kaolinite in Reaction 12-8.

Calcite neutralizes acid by dissolving and complexing with hydrogen ion to form bicarbonate and carbonic acid. These inorganic carbon species remove free hydrogen from the solution, thereby lowering its acidity. The calcite neutralizing reactions can be shown as:

$$CaCO_3 + H^+ \rightarrow Ca^{2+} + HCO_3^- \tag{12-9}$$

$$HCO_3^- + H^+ \rightarrow H_2CO_3 \tag{12-10}$$

The presence of carbonic acid (H_2CO_3) in solution provides a future source of acid (H^+) to solution; consequently, its formation is not a permanent reduction in acidity. However, if excess dissolved CO_2 gas can be released from solution by degasing, then the acidity can be permanently lowered by the following reaction:

$$H_2CO_3 \rightarrow CO_2(aq) + H_2O \rightarrow CO_2(gas)\uparrow$$

At a site in Tennessee where acid mine surface water contacts limestone, Webb and Sasowsky[7] found that dissociation of carbonic acid and loss of CO_2 gas to the atmosphere accounted for about 30% of the neutralization. The remainder was attributed to dissolution of carbonate minerals.

In addition to the solid phases of carbonate in the subsurface, dissolved carbonate and bicarbonate provide neutralizing capacity to the system by reactions such as 12-10. However, the amount of neutralizing capacity provided by dissolved species under typical groundwater concentrations is usually small compared with solid-phase carbonates even where the minerals are present in very small amounts. For example, 100 mg/L HCO_3^- can neutralize about 1.6 milliequivalent of H^+, whereas 0.01 wt % calcite (100 ppm) in contact with 1 liter of groundwater can neutralize 8 milliequivalents of H^+.

In areas where carbonate minerals are absent and the dissolved inorganic carbon content is low, the neutralizing capacity of dissolved organic carbon in solution may be important. Litaor and Thurman[8] showed that, for various vegetation types in an alpine watershed in Colorado, organic acids that deprotonate in the pH range of 4 to 4.5 (e.g., carboxylic functional groups) provide from 20 to 52% of the acid-neutralizing capacity of the soil solution.

Clays minerals can also provide neutralizing capacity to the system, but generally at a lower pH than carbonate minerals.[6] If all of the carbonate minerals have been dissolved by acid or these minerals were absent from the solid phase, then the clay minerals can consume hydrogen ions as they dissolve. The dissolution of the clay mineral kaolinite shows this neutralizing ability:

$$Al_2Si_2O_5(OH)_4 + 6H^+ \rightarrow Al^{3+} + 2H_4SiO_4^0 + H_2O \tag{12-11}$$

If the dissolved Al^{3+} is allowed to precipitate as gibbsite, this neutralizing mechanism is lost because an equal amount of hydrogen will be released into solution:

$$2Al^{3+} + 6H_2O \rightarrow 2Al(OH)_3 + 6H^+ \tag{12-12}$$

A final neutralizing source in the subsurface is the base cations (Ca^{2+}, Mg^{2+}, Na^+, and K^+) present on the exchange sites of clays and organic matter. Hydrogen ion added to the system will compete for these cation exchange sites and will be partially sorbed onto these sites in accordance with the solution concentration of the competing cations. The reaction with exchangeable calcium can be represented as follows:

$$clay-(Ca^{2+})_{0.5} + H^+ \rightarrow clay-H^+ + 1/2\,Ca^{2+} \tag{12-13}$$

For typical groundwater with major cation concentrations on the order of 1 millimol/L, hydrogen ions do not successfully compete for the exchange sites until the pH is less than about 5 because of the low hydrogen ion activity (<0.01 millimol/L) at pH values > 5. Therefore it is only under fairly acidic conditions that the base exchange mechanism affects the acidity of the water. The capacity of this mechanism is a function of the cation exchange capacity of the system. Moss and Edmunds[9] describe two relatively pristine sites in the British West Midlands that have acidic soil solutions (pH 4.0 and 4.5) and lack carbonate minerals. They attribute the effective acid-neutralizing capacity in the unsaturated zone to dissolution of K-feldspar and cation exchange reactions with Ca^{2+} and Mg^{2+} on the clay mineral surfaces. They believe an important mechanism of neutralization at this site is cation exchange and surface binding of the acid-forming cation Al^{3+}:

$$clay-Ca_3 + 2Al^{3+} \rightarrow clay-Al_2 + 3Ca^{2+} \tag{12-14}$$

Removing Al^{3+} from solution minimizes the formation of gibbsite and the release of hydrogen ions (Equation 12-12).

The neutralization reactions described above occur under primarily oxidizing conditions, which produced the original instability for the sulfide mineral. Neutralization of acid and immobilization of groundwater contaminants can also occur under reducing conditions. The following overall reaction has been shown to occur in lake sediment with reactive organic matter:[10-12]

$$4Fe(OH)_3 + 4SO_4^{2-} + 9CH_2O \rightarrow 4FeS + 8HCO_3^- + 11H_2O + CO_2 \qquad (12\text{-}15)$$

In this reaction iron and sulfate are reduced by the organic matter CH_2O and precipitated as the iron monosulfide FeS. The reaction is catalyzed by sulfate-reducing bacteria. The generation of bicarbonate provides acid-neutralizing capacity to the system. This neutralization mechanism will only be permanent if the FeS is not allowed to oxidize in the future, such as by burial in lake sediments. This neutralization may also occur in an aquifer environment where the acidic plume contacts a zone of reactive organic matter that has sufficient reducing capacity to lower the Eh to the stability field of the sulfide minerals (Figure 12-1).

In summary, the most common reactive neutralizing agents in surface water or groundwater systems are dissolved bicarbonate and carbonate; however, the neutralizing capacity of this source is very limited. Carbonate minerals present in the system provide the next most reactive phase, and their neutralizing capacity is a function of their concentration and availability to the acidic solution. Rapid precipitation of ferrihydrite on neutralization by carbonate minerals may coat the carbonate solids, reducing their reactivity, or may coat the sulfide minerals, making them less susceptible to continued oxidation.[7] If neither dissolved or solid inorganic carbon is present in the system, then organic acids may provide some limited neutralizing capacity proportional to their dissolved concentration. As the pH of the system decreases, cation exchange, as well as the dissolution of clay minerals, may consume acidity. Under reducing conditions, bacterial sulfate reduction can provide alkalinity to the system by the generation of bicarbonate. Residual neutralizing capacity will be provided by the primary silicate and aluminosilicate minerals, but these reactions are slow on the time frame of typical groundwater residence and are not expected to be rapid neutralizers of acidic solutions. These general acid-neutralizing reactions are not limited to the example of pyrite oxidation. Any acidic source will tend to be neutralized by one or more of these mechanisms.

12.2. ACIDIC ENVIRONMENTS

Sulfide mineral oxidation and the production of acidic conditions, as described above, occur in a variety of man-made environments and in certain unperturbed systems. Acidic conditions created by human activities include acid mine drainage; storage piles of ore, waste rock, and tailings; and lakes associated with excavation of ore deposits. The characteristics of man-made and naturally acidic environments are described in this section.

Naturally Acidic Conditions
The natural occurrence of highly acidic conditions due to, for example, pyrite oxidation is somewhat rare because the rate of acid generation can be balanced by neutralization reactions. Water and oxygen in the unsaturated zone will oxidize pyrite, but the solution may in turn be neutralized as it flows through the unsaturated and saturated zones on its path to a discharge point. If the neutralizing capacity of the material along the flowpath has been consumed or the rate of neutralization is too slow for the residence time of the groundwater, then the discharge water may remain acidic, but this is likely only for a very short flowpath in highly weathered material. The rate of exposure of fresh pyrite and other sulfide minerals to near earth surface oxidizing conditions, hence the rate of acid generation, will be governed by the weathering rate, which is generally very slow for chemical weathering. This will provide time for neutralizing reactions to occur as the landscape

naturally weathers and slowly exposes new sulfide minerals to oxidizing conditions. Mechanical weathering such as mass wasting, which is primarily important in areas of high relief, may rapidly expose acid-producing minerals to earth surface conditions. In these cases localized zones of acidity may be created and surface waters may be acidic because of the relatively limited opportunities for neutralization by water/rock interactions. The Summitville (Colorado) area contains a number of drainages that appear to have been impacted by acid conditions produced by the natural process of sulfide mineral weathering.[13,14]

Acid Mine Drainage

One of the most common occurrences of acidic conditions is associated with the mining of coal and base or precious metals. Pyrite occurs with the ore material because the environment in which the coal or metals were deposited is reducing and is conducive to the formation of pyrite and other sulfide minerals along with the ore. Mining operations expose sulfide minerals in the host rock to atmospheric oxygen either in the underground workings or in surface spoils. In the case of coal deposits, pyrite is a minor component of the organic ore material; however, for metal deposits, pyrite may be present at high concentrations relative to the metal being extracted. In either case it does not require large amounts of pyrite to create highly acidic conditions.

The occurrence of acid mine drainage in the copper-zinc West Shasta Mining District of northern California has been extensively studied because of its impact on nearby surface water. Nordstrom et al.[15] characterized mine effluent and its effect on Boulder Creek and downstream Spring Creek. They measured pH values as low as 1.1 in the mine effluent with the pH in the creeks increasing to 2.9 over a distance of 14 km from the point of effluent discharge. Iron and sulfate concentrations exceeded 10,000 and 60,000 mg/L, respectively, in the effluent and decreased downstream to less than 1000 mg/L due to dilution and probable precipitation of iron and sulfate minerals, such as ferrihydrite ($Fe[OH]_3$) and jarosite ($KFe_3[SO_4]_2[OH]_6$). Along the surface water flowpath, aluminum concentrations decrease from 1400 to 47 mg/L, partially due perhaps to the precipitation of aluminum sulfate minerals such as jurbanite ($Al[SO_4][OH]\cdot 5H_2O$).

The effect of acid mine drainage on West Squaw Creek, which is north of Boulder and Spring Creeks and drains into Shasta Lake, has been studied by Filipek et al.[16] They also found low pH and high metal and sulfate concentrations in the creek several miles downgradient of mine sites along West Squaw Creek; however, the pH was generally higher (2.5 to 5.5) than at Boulder and Spring Creeks and the dissolved concentrations were all generally less than 1000 mg/L. They observed an amorphous orange precipitate associated with algae in the stream beds, which they believed was ferric hydroxide or a mixture of ferric hydroxide and jarosite. Elevated arsenic concentrations (>10 μg/L) were reduced to below detection level (1 μg/L) a short distance from the source as the arsenic was removed from solution probably by adsorption onto the precipitating iron minerals. These two sites in the West Shasta Mining District illustrate the high mobility of an acidic, metal-rich solution in an environment with low-neutralizing ability.

Strip mining of coal also has the potential for producing acidic conditions if the coal contains sulfide minerals and has a low-neutralizing capacity. Karathanasis et al.[17] sampled soil solutions and surface water from unreclaimed and partially reclaimed watersheds that had been strip-mined in Kentucky. They found that after almost 20 years since mining ceased, the pH of the surface water and soil was still depressed (2.7 to 4.3) and that water samples with the lowest pH values (<3) had very high concentrations of Al and Fe (50 to 2000 mg/L) and sulfate (400 to 13,000 mg/L). They believed that the dissolved Al concentration may be limited by the formation of a jurbanite-like mineral ($Al[SO_4][OH]_6$) and alunite ($KAl_3[SO_4]_2[OH]_6$). The dissolved iron level appeared to be limited by a mineral with the stoichiometry of jarosite ($KFe_3[SO_4]_2[OH]_6$), but with a much higher solubility.

Acid mine drainage that has had the opportunity to come into intimate contact with neutralizing minerals, such as along the flow path of an aquifer, is not as likely to remain acidic as are surface waters. Stollenwerk[18] investigated the contact of acidic water with alluvial aquifer material in the Globe-Miami copper mining district of Arizona. The most contaminated groundwater near the source had a pH of 3.3 and dissolved concentrations of 9600 mg/L SO_4^{2-}, 2800 mg/L Fe, 300 mg/L Al, and 200 mg/L copper. Uncontaminated parts of the alluvium contained about 0.3 weight % calcite, which provides the primary neutralizing capacity in the aquifer system. Near the source, the calcite had dissolved allowing a low-pH plume to form in the aquifer. Over a distance of 10 km from the source, the pH of the groundwater increased slowly from 3.3 to about 4.3 at which point it increased relatively rapidly over the next two kilometers to 7. The neutralized groundwater has very low Fe, Al, and Cu concentrations (<1 mg/L), while the sulfate level (2000 mg/L) remains well above the background concentration (70 mg/L). The decrease in Fe concentration is attributed to the low solubility of $Fe(OH)_3$ at neutral pH. Dissolved aluminum levels appeared to be limited by the precipitation of amorphous $Al(OH)_3$ at pH > 4.7 and $AlOHSO_4$ at pH < 4.7. The concentration of sulfate is controlled by the relatively soluble mineral gypsum, which explains the high level even in the neutralized water. Attenuation of the metals Cu, Co, Mn, Ni, and Zn along the groundwater flowpath appeared to be controlled by adsorption onto ferrihydrite.

The least likely setting for the occurrence of acid mine drainage is a mine site in carbonate terrain; however, sulfide oxidation can still produce groundwater contamination. The zinc-lead mines near Shullsburg, Wisconsin, are located in a carbonate aquifer. This aquifer was dewatered during mining allowing contact of the sulfide minerals (pyrite, galena [PbS], and sphalerite [ZnS]) with atmospheric air. Toran[19] found that the carbonate minerals dolomite and calcite effectively neutralized acid production and that siderite ($FeCO_3$) and ferrihydrite limited iron concentration to an acceptable level. However, the sulfate level in groundwater allowed to reenter the mine since shutdown of the main mine working in 1979 is elevated (96 to 3800 mg/L) within a half mile of the mine. Calculation of the gypsum saturation indices for groundwater samples showed it to be undersaturated in most cases despite high dissolved levels of calcium and sulfate. If gypsum is present as a byproduct of sulfide mineral oxidation in the mine workings, then the dissolved sulfate from dissolution of this gypsum into the groundwater is diluted by mixing with freshwater a short distance from the workings. Toran concluded that the sulfate contamination is localized and will decrease with time by continued dissolution and removal of the source and dilution of the contaminated groundwater.

Ore Storage, Mine Waste and Tailings Piles

The presence of sulfide-containing ore, mine waste, or mill tailings on the surface of the earth creates a localized environment with a high potential for acid production. Oxidizing water percolating through the material dissolves the sulfide minerals, creating acidic conditions unless the material also contains neutralizing solids. In the long term the overall rate of oxidation will be controlled by the delivery of oxygen to the sulfide minerals. Oxygen can be transported by diffusion in the gas phase for unsaturated conditions and dissolved in the solution phase for saturated conditions.[20] The progression of acidic conditions in the subsurface beneath the pile will be a function of the flux of acidity from the pile and the neutralizing capacity of the native material.

Blowes and Jambor[21] evaluated the porewater geochemistry of sulfide tailings at a former Zn–Cu mine in Quebec. They described three zones within the tailings. Oxidation has depleted sulfides in the upper zone, which is generally less than 0.5 m thick. In the intermediate zone, sulfide oxidation and neutralization reactions continue. Where the water table is shallow leading to low O_2 diffusivity, the intermediate zone is thin to nonexistent, but where the tailings are coarser and the water table deeper, the intermediate zone can be up to 1 m thick. Below the intermediate zone is an unoxidized zone beneath the depth of oxygen penetration. Because the average sulfide content of the unoxidized tailings material is 50 wt % and the carbonate content is 1.5 wt %, the tailings have a much greater acid generating ability than neutralizing capacity. Geochemical processes other than disso-lution of carbonate minerals that can neutralize the acid are dissolution of aluminosilicate minerals, dissolution of Fe- and Al-hydroxide minerals, and H^+ adsorption on mineral surfaces. It is calculated that the acid generated in oxidizing 1 m of tailings requires a 25-m flowpath to be neutralized by these reactions. For this reason the acid plume can move out of the zone of active acid generation into the lower tailings and the native material below the pile.

The calculation of mineral saturation indices allowed Blowes and Jambor to identify possible mineral phases limiting solution composition in this tailings pile. Goethite (α-FeOOH), lepidocrocite (γ-FeOOH), ferrihydrite ($Fe[OH]_3$), jarosite ($KFe_3[SO_4]_2[OH]_6$), and gypsum ($CaSO_4 \cdot 2H_2O$) limit the solution concentrations of the major ions, Fe^{2+}, SO_4^{2-}, Ca^{2+}, Na^+, and K^+ in the pore water. Lead may be limited by anglesite ($PbSO_4$), and Cr may be limited by the precipitation of amorphous $Cr(OH)_3$, while Cu is precipitated as covellite (CuS) below the zone of active oxidation. There do not appear to be any mineral solubility limits on Co, Mn, Ni, or Zn concentrations; therefore it is likely that the migration of these metals through the tailings is primarily affected by adsorption/desorption processes. Given the current rate of oxidation of this tailings pile and the amount of sulfide minerals present, it is calculated that these tailings will continue to generate acid for several centuries unless the oxygen source is cut off or additional neutralizing agents are added.

The precipitation of secondary minerals at different "fronts" in a tailings pile can affect the movement of porewater by forming horizontal barriers to vertical flow. Blowes et al.[22] discuss the occurrence of a discontinuous, cemented layer (hardpan) at the Quebec tailings pile described above. This layer is 1 to 5 cm thick and occurs at the depth of active sulfide mineral oxidation. Cementation is produced by Fe(III) minerals, principally goethite, lepi-docrocite, ferrihydrite, and jarosite. At a separate, inactive, sulfide tailings site in New Brunswick, they describe another type of hardpan that is 10 to 15 cm thick and continuous and occurs 20 to 30 cm below the depth of active oxidation at the first appearance of calcite. In this case cementation is caused by the precipitation of gypsum and Fe(II) solid phases, principally, melanterite ($FeSO_4 \cdot 7H_2O$). The lack of precipitation of the Fe(III) oxide and sulfate minerals at this site is attributed to a higher sulfide content in the tailings and a lower pH, which increases the solubility of the Fe(III) minerals, minimizing their formation. Concentrations of Fe(II) and sulfate increase until melanterite becomes supersaturated. Gypsum also precipitates when calcite is encountered because the dissolution of calcite releases calcium that can combine with sulfate to form gypsum. The presence of these hardpan layers in the tailings piles are viewed as a benefit because they limit the movement of solution and gases through the reactive tailings material.

Pyrite is a ubiquitous constituent in coal. In a review of water-quality issues associated with coal storage, Davis and Boegly[23] compile data on leachate from coal storage piles. They found pH values as low as 2.1, iron concentrations as high as 93,000 mg/L, and sulfate levels to 22,000 mg/L. Concentrations of Zn, Pb, As, Cu, Cr, and Al exceeded 1 mg/L in the majority of the tabulated leachate compositions. Helz et al.[24] identified the minerals melanterite ($FeSO_4 \cdot 7H_2O$) and ferrihydrite as concentration-limiting minerals in a leaching study of Appalachian coals. These minerals are fairly soluble under the conditions of the coal pile and do not limit solution concentrations to low levels. They recommend evaluating the addition of limestone to coal piles for neutralization purposes in cases where the leachate is causing environmental problems.

Acidic tailings piles have also been produced by processing of ore material with sulfuric acid. This is one of the most common methods of extracting uranium from its ore. The environmental effect of land disposal of the acidic tailings has received much scrutiny because of the potential for leaching and transport of metals and radionuclides from these tailings.[25-28] Because the acidity of the tailings porewater in this case is not generally produced by oxidation of sulfide minerals in the tailings, the acid generating ability of the solid phase is limited and the subsurface soils have a better chance of neutralizing the leachate. At many of these sites it has been found that the pH of the leachate is raised from a value of less than 2 in the tailings to near neutral by contact with less than one meter of subsurface native material. Most of these sites are in the arid to semiarid western United States where calcite is a common primary or secondary mineral in the soil. Calcite neutralizes the acid, leading to the precipitation of other minerals such as ferrihydrite, gypsum, $Al(OH)_3$, and $AlOHSO_4$ (at low pH). Other metal contaminants and radionuclides associated with the leachate appear to be adsorbed or coprecipitated with these secondary minerals. The most mobile contaminant at these sites is usually sulfate, which is present at elevated concentrations in groundwater because of the relatively high solubility of gypsum.

Acid Lakes

Lakes and other surface water bodies may become acidic because of the inflow of acid waters from mine workings, as in the case of mine pit lakes, or because of acid precipitation in regions of elevated atmospheric content of acid-producing gases, especially SO_x and NO_x. The geochemistry of these two environments with respect to acid generation and neutralization is discussed in this section.

Mine Pit Lakes

Open-pit, hard-rock mining for metals commonly extends below the water table and requires dewatering during the mining process. At the cessation of mining, the pit fills with water to the level of the water table creating a lake that may be very large (containing >10,000 acre-feet of water). Water-quality issues associated with these lakes have been reviewed by Miller et al.[29] Because of the presence of sulfide minerals associated with the ore material and the exposure of these minerals to atmospheric conditions during mining, water draining into the pit has the potential to be acidic. The deeper the pit, the more likely that reducing conditions will be encountered and pyrite present in the host rock. Data compiled by Miller et al. for the Berkeley Pit (Butte, Montana) and the Liberty Pit (White Pine County, Nevada) show pH values of 2.8 and 3.21, respectively; Fe, Cu, and Zn concentrations greater than 50 mg/L; and sulfate greater than 3500 mg/L. Saturation index calculations for water in the Berkeley Pit showed that the minerals ferrihydrite, gypsum, jurbanite ($AlOHSO_4$), and jarosite ($KFe_3[SO_4]_2[OH]_6$) provide solubility limits for the dissolved constituents under acidic conditions.[30]

In cases where the capacity of the system is sufficient to neutralize acid generation, the pH values are not depressed. For example, in the lakes at the Kimberly Pit (White Pine County, Nevada) and the South Pit, Getchell Mine (northern Nevada), the pH values are 7.61 and 5.96, respectively.[29] Elevated sulfate concentrations (>1000 mg/L) suggest that sulfide mineral oxidation is occurring at these sites, but, in general, metals concentrations are not significantly elevated because of the low solubility of their minerals and/or high-adsorption affinity at these near-neutral pH values. However, arsenic is of potential concern in the more neutral lake environments because it is less strongly adsorbed at high pH (>6.5) than under acidic conditions. At the Getchell Mine the arsenic concentration is 40 times greater in the higher pH (7.67) North Pit lake than in the South Pit lake (pH = 5.96).

Acid-Impacted Natural Lakes

Lakes downwind of coal-fired power plants and industries that burn sulfur-containing coal can become acidified by the introduction of acid-producing gases (principally SO_x) into their waters. Oxidation of these gases produces sulfuric acid. In the general case for SO_2 the reaction is as follows:

$$SO_2 + 1/2\,O_2 + H_2O \rightarrow SO_4^{2-} + 2H^+ \qquad (12\text{-}16)$$

In areas where the neutralizing capacity of the system is sufficient, the pH of the lake will not be depressed. However, in the northeastern United States where the natural neutralization capacity is low because of the lack of carbonate minerals and relatively high natural organic acid content of runoff, this additional acid from atmospheric deposition may depress the pH of the lake. For example, at Woods Lake (New York) the acidic precipitation has a mean pH of 3.96 and the pH of the lake due to this anthropogenic acidic deposition is 5.0.[31]

The fate of sulfate in surface water may also have an impact on lake acidity. Sulfate may be tied up in the sediments as organic S compounds or reduced inorganic S. In a study of eight acid lakes in Quebec, Carignan and Tessier[11] found that the majority of the anthropogenically derived excess S was tied up in the sediments as reduced inorganic iron minerals such as FeS and pyrite. As shown by Reaction 12-15, the reduction of Fe(III) and SO_4^{2-} in the sediments by organic matter, as facilitated by sulfate-reducing bacteria, produces not only the iron sulfide minerals but bicarbonate in solution. Additional sulfate-reducing reactions postulated by Carignan and Tessier are as follows:

$$8CH_2O + 4SO_4^{2-} + 2O_2 \rightarrow 4S^0(s) + 8HCO_3^- + 4H_2O \qquad (12\text{-}17)$$

$$17CH_2O + 8SO_4^{2-} + 4Fe(OH)_3(s) + 2O_2 \rightarrow 4FeS_2(s) + 16HCO_3^- + 15H_2O + CO_2 \quad (12\text{-}18)$$

In each case the reduction of one mole of sulfate produces two moles of alkalinity as represented by HCO_3^-. The bicarbonate produced in the sediments can participate in acid neutralization by the following reaction:

$$HCO_3^- + H^+ \rightarrow H_2CO_3 \qquad (12\text{-}19)$$

Herlihy et al.[10] also showed that for a lake in Virginia that received acid mine drainage, the major sink for iron and sulfate was bacterial sulfate reduction and the precipitation of FeS in sediments. As long as the sulfide minerals are stable in the reduced condition of the sediment and are not reoxidized, this mechanism of creating neutralizing capacity will be effective.

In some sediments Carignan and Tessier found that the Fe(III) reserve available to form iron sulfide minerals was nearly or completely exhausted. These lakes will have a lower capacity to neutralize acid input through the process of sulfate reduction and bicarbonate production. In lakes with less available iron it is more likely that the sulfur in sediments will be contained in organic S compounds.

The presence of acidic conditions in surface water should not always be attributed to anthropogenic pollutants. In a study of the factors controlling acidity in the River Lillån, northern Sweden (pH = 5.18), Jansson and Ivarsson[32] found that the acidity in 38 small tributary streams to the river was caused principally by naturally occurring organic acids. The dissolved organic carbon in most of the streams varied between 10 and 30 mg/L. During periods of rainstorms the precipitation flushed organic acids from the forest soils and lowered the pH in the streams from 6.5–7.0 to 4.5–5.0. The movement of humic substance from soil produced the majority of the acid conditions in the river. In this area of Sweden, which is considered susceptible to the impact of acid precipitation, natural sources of acidity may also be important if not dominant.

12.3. SAMPLING CONSIDERATIONS

In some ways the collection of water samples from an acidic site impacted by sulfide oxidation is simpler than collecting groundwater at neutral or alkaline pH values. There will be little dissolved inorganic carbon, CO_2 gas, or alkalinity in the water; therefore loss of carbonate on degasing of CO_2 will not be an issue. Also, the low pH of the water will aid in preservation of dissolved metals. The laboratory should be alerted that the samples have high acidity and dissolved metal levels. If acidic conditions are a consequence of high concentrations of organic acids in a weakly buffered groundwater, then the dissolved organic carbon concentration and identity of the organic acids must be determined. The noncarbonate alkalinity attributed to these organic acids must be subtracted from the total measured alkalinity before calculating the carbonate alkalinity and concentrations of bicarbonate and carbonate in these waters. For water samples collected from the zone of neutralization in which carbonate minerals are the neutralizing agent, loss of CO_2 gas on sampling and its effect on the pH of the water sample will need to be considered (see Sections 9.1 and 9.5).

At acid-impacted sites, the major dissolved constituents are commonly much different than typically present in neutral to alkaline groundwater. Iron and aluminum are often the metals present in highest concentration with levels exceeding 1000 mg/L when the pH is less than 3. Depending on the ore material, transition metals such as Pb, Ni, Cr, Zn, Cu, and Co may also be present at concentrations greater than 10 mg/L. It is important that the sampling program require analysis for these metals, in addition to the more common major groundwater cations Ca, Mg, Na, and K. Sulfate is generally the major anion at sulfide-oxidized acid sites; however, at mill tailings facilities other acids such as HNO_3, HCl, and HF may have been used for ore processing, and the anions of these acids would need to be measured in the water samples.

Waters with low pH and relatively high iron concentration appear to be one of the few solutions where the platinum-electrode field Eh is representative of the system Eh. These systems are well poised because of the high concentrations of both redox-active iron species Fe(II) and Fe(III). Nordstrom et al.[15] found that in surface waters impacted by acid mine drainage in northern California, the calculated Eh value from the measured concentrations of Fe(II) and Fe(III) corresponded closely with the field measured value. Davis et al.[33] found that, for an acid mine drainage environment in Colorado, the field Eh matched the Fe(II):Fe(III) calculated Eh at pH values less than 3.0 and matched the Eh calculated from the Fe(II):Fe(OH)$_3$(s) reaction at pH values greater than 3.0. They proposed that a minimum of 1 mg/L of total dissolved iron was necessary to poise the Eh. Davis and Ashenberg[30] were also generally successful at comparing the calculated Fe(II):Fe(III) Eh value with the field Eh for the acid water in the Berkeley Pit (Montana). They concluded that the iron redox couple can poise the Eh if both iron species (Fe[II] and Fe[III]) are present in solution at concentrations greater than 0.25 mg/L and the minor of the two species is at least 5% of the total dissolved iron concentration.

To evaluate contaminant migration and design an effective remediation system, analysis of the vadose zone and aquifer solids is also essential. The neutralizing capacity of the system must be determined to estimate the rate of movement of the acid front. Neutralization will be due principally to the presence of cation exchange surfaces, carbonate minerals, reactive metal (Fe, Mn, and Al) oxyhydroxides, and clay minerals. The identity, concentration, and mode of occurrence of reaction products that form as the acid water attempts to equilibrate with the subsurface environment should also determined. Secondary minerals such as iron, aluminum, and calcium sulfates may create a long-term source of acid or sulfate contamination in the subsurface. The precipitation of the secondary minerals may also armor some of the neutralizing solids, thereby decreasing the effective neutralizing capacity of the system, allowing the acid front to move farther and faster than anticipated.[28]

As with the design of any sampling program, a conceptual model of the geochemical interactions at the site will allow the investigator to focus on the potentially reactive components of the system and characterize its most important features.

12.4. GEOCHEMICAL MODELING

Simulating the water/rock interactions that accompany acid generation and neutralization is popular and useful because the system is so geochemically reactive and such a wide variety of processes occur simultaneously. The oxidation of pyrite catalyzed by bacterial mediation is a relatively fast process that produces a strong acid, which enhances weathering and other neutralization reactions. Many mineral phases may be dissolving and precipitating simultaneously during neutralization, while adsorption/desorption of H^+ and metals onto preexisting mineral surfaces and newly formed precipitates (primarily ferric and aluminum hydroxides) also affects contaminant mobility. In addition, many speciation reactions are pH-dependent, and they effect solution composition over the wide range of pH (<2 to >8) possible during acidification and neutralization.

Geochemical modeling calculations are often used in these environments to calculate mineral saturation indices and identify minerals that might be forming and limiting solution concentration for their constituents. The primary minerals that have been identified in the more acid environments are melanterite ($FeSO_4 \cdot 7H_2O$), ferrihydrite ($Fe[OH]_3$), jarosite ($KFe_3[SO_4]_2[OH]_6$), jurbanite, ($Al[SO_4][OH] \cdot 5H_2O$), alunite ($KAl_3[SO_4]_2[OH]_6$), and gypsum ($CaSO_4 \cdot 2H_2O$). In the less acid conditions as neutralization raises the pH, the precipitating minerals have been identified as ferrihydrite ($Fe[OH]_3$), goethite (α-FeOOH), lepidocrocite (γ-FeOOH), gibbsite ($Al[OH]_3$), and gypsum ($CaSO_4 \cdot 2H_2O$). Under reducing conditions that might occur in the sediment of an acid lake with reactive organic matter, the minerals mackinawite (FeS) and pyrite (FeS_2) have been shown to be capable of precipitating.

Because the low pH environment may mobilize many trace metals such as Pb, Cu, Zn, Sr, Cr, and As that may not be present in high enough solution concentration to precipitate a mineral, adsorption is a necessary geochemical process to model for metals at acid sites. The adsorption modeling capabilities of MINTEQA2 have been used to simulate the mobility of these metals as an acid plume is neutralized on contact with native material.[18,27,28]

Because of the potential for acidic conditions to develop in mine pit lakes, the state of Nevada requires mining companies to use geochemical modeling to predict future water quality in the lake and help estimate environmental risk.[29] For this purpose the mixing capability of the PHREEQE code could be used to evaluate the various inputs to the lake and precipitate or dissolve reactive minerals present in the system. Also, the PHREEQE titration ability has been used for the Berkeley Pit (Montana) to evaluate the feasibility of using basic tailings as an amendment to the acidic pit waters to raise the pH and reduce metals concentrations.[30] Modeling the neutralization processes provided information on the required amount of alkaline fluid to add to the pit water and the optimum pH to immobilize metals. Because most metals such as the cations Pb^{2+}, Zn^{2+}, and Ni^+ are more strongly immobilized under alkaline conditions and because semimetals such as the predominantly anionic As and Se are more strongly immobilized under acid conditions, the optimum pH will be a function of the specific contaminants of concern at a site, the solution composition, and the surface complexation properties of the available adsorbents.

12.5. CONTAMINANT TRANSPORT

Acidic environments, particularly at pH values less than 3, are very conducive to the mobility of contaminants at very high concentrations. As described in Section 12.2, sulfate solution concentrations can be much greater than 10,000 mg/L, Fe and Al values greater than 1000 mg/L are common and the "trace" metals (Pb, Cr, Cu, Zn, Ni, etc.) often exceed 10 mg/L.

The mobility of these constituents away from the source of acidity is primarily a function of the total acidity of the solution and the acid-neutralizing capacity of the material the solution contacts. The acidity of the solution is partially due to the activity of hydrogen; however, a much larger component is generally due to dissolved Fe and Al. As the pH of the solution is raised by reactions with the solid phase, Fe and Al minerals become less soluble and precipitate. The reactions to form ferrihydrite and gibbsite can be represented as follows:

$$Fe^{3+} + 3H_2O \rightarrow Fe(OH)_3 + 3H^+ \tag{12-20}$$

$$Al^{3+} + 3H_2O \rightarrow Al(OH)_3 + 3H^+ \tag{12-21}$$

For each mole of Fe and Al precipitated, 3 moles of hydrogen are produced. If the original Fe and Al concentrations in the acidic water are each 1000 mg/L, then precipitation of ferrihydrite would create 54 mmol/L H$^+$ and precipitation of gibbsite would produce 110 mmol/L H$^+$. This is much greater acidity than the 10 mmol/L H$^+$ provided by the solution concentration of hydrogen at a pH of 2. As a consequence, the pH plume and its dissolved constituents will be more mobile in an acidic solution with high concentrations of Fe and Al than in a plume without these metals.

In the discussion of uranium mill tailings above, the rapid neutralization and immobilization of contaminants was attributed to the fact that acidity was limited primarily to the solution phase and that the tailings solids did not provide significant additional acidity through weathering processes such as sulfide oxidation. At other sites where sulfide-bearing material has been stockpiled or disposed on the earth's surface, the acid-generating capacity can be much higher. If this material has been exposed for long periods, the resultant groundwater plume developed from these materials can extend for several miles downgradient of the source. The plume will typically consist of a zone of low-pH water where metals concentrations can be elevated and the neutralizing capacity of the aquifer has been consumed or is no longer in direct contact with the groundwater (Figure 12-2). (It has been shown in column experiments of acid neutralization that the precipitation of ferrihydrite can coat the reactive calcite and inhibit further neutralization.[28]) The low-pH region in the aquifer ends at the neutralization zone where the aquifer has the capacity to consume free hydrogen in solution and hydrogen produced by precipitation of metal hydroxides. Metals are expected to precipitate in the neutralization zone because of the lower solubility of their solids and the strong adsorption of trace metals on the precipitating Fe and Al hydroxides. The neutralization zone moves downgradient in the aquifer in response to the addition of acidic solution from the source. The rate of movement is a function of the flux of acidity from the source and the acid-neutralizing capacity of the material along the flow path, both of which may be quite variable.

The major contaminant that can move beyond the neutralization zone is sulfate. Its concentration in solution is not directly pH dependent; therefore neutralization will not inhibit its mobility as it does the metals. The total dissolved concentration of sulfate is usually limited by the solubility of gypsum; however, the solubility of this mineral can vary by orders of magnitude depending on the other constituents in solution. In dilute solutions the expected concentration of sulfate in equilibrium with gypsum is on the order of 2,000 mg/L, but in acidic solutions with high concentrations of divalent (Ca^{2+}, Mg^{2+}, Fe^{2+}) and trivalent (Fe^{3+}, Al^{3+}) cations, sulfate forms strong solution complexes that increase the solubility of gypsum. As a consequence, the concentration of sulfate in equilibrium with gypsum in the acid zone may be greater than 50,000 mg/L. In the neutralization zone, the solubility of gypsum will decrease as the cations are removed from solution; however, its solubility can easily remain high enough that sulfate will exceed the EPA secondary drinking

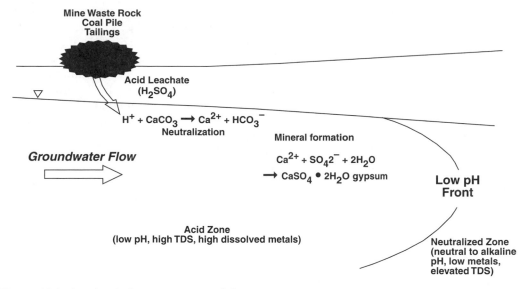

Figure 12-2 Geochemical processes — acid drainage.

water standard of 250 mg/L by a factor of ten. Because of the relative mobility of sulfate compared with the other constituents in this type of acid plume, sulfate is the best indicator of the first arrival of contaminants from a sulfide source.

12.6. REMEDIATION

Before remediation of an acid-contaminated aquifer is attempted, the source must be eliminated. The addition of limestone or other carbonate mineral to an acid-generating source is a common practice. The carbonate serves to neutralize any acid produced and keeps the pH high, thereby slowing acid generation. Webb and Sasowsky[7] showed that allowing degasing of CO_2 produced during the neutralization process with carbonate solids will further eliminate acid from the system. A further possible benefit of using a rapid neutralizer such as carbonate minerals was identified by Nicholson et al.[34] who found that ferrihydrite may precipitate on the pyrite during neutralization, protecting it from further oxidation. Conversely, if ferrihydrite precipitates on the carbonate minerals it will inhibit neutralization.[28] Webb and Sasowsky[7] suggest treating acid mine drainage in "anoxic mine drains" to prevent ferrihydrite from precipitating and coating the added carbonate solid. However, this would not prevent the migration of iron-rich water and would curtail the adsorption of trace metals onto the precipitating ferrihydrite.

Rather than treating the acid in the source area with a neutralizing agent after the acid has formed, an alternative exists in minimizing the generation of acid. Because sulfide oxidation is bacterial mediated, principally by *Thiobacilli*, use of bactericides has been recommended to reduce acid production. Kleinmann et al.[4] recommend the use of a simple anionic detergent (sodium lauryl sulfate) that they found significantly lowered acidification in fresh coal storage piles. The inhibitor can be sprayed on for short-term benefit or added in a controlled release compound that might provide benefit for several years.

If an acidic plume has entered an aquifer, then water/rock interactions will influence remediation design and the degree of remediation required. The relatively high mobility of the contaminants in an acid plume facilitates their removal by the pump-and-treat method; however, cation exchange and mineral dissolution may extend the time required to achieve cleanup. At low pH and high dissolved Fe and Al concentrations, the cation exchange sites on the surfaces of minerals become loaded with H^+, Fe^{2+}, Fe^{3+}, and Al^{3+}, displacing the typical groundwater major cations (primarily Ca^{2+} and Mg^{2+}). As fresh groundwater moves into the zone of contamination during pump-and-treat remediation, Ca^{2+} and Mg^{2+} repopulate the exchange sites and release H^+, Fe, and Al to solution. The release of H^+ and the acid-forming metals to solution prolongs the acid conditions, requiring additional pumping to achieve remediation. By proper characterization of the acid zone, its acid-forming potential can be estimated and compared with the neutralizing capacity of the fresh groundwater. This comparison will aid in estimating the volume of water required to neutralize the system. Remediation may be enhanced by adding a neutralizing agent such as HCO_3^- or CO_3^{2-} to the water flowing through the acid zone.

The presence of iron, aluminum, and calcium sulfate minerals such as jarosite, jurbanite, and gypsum precipitated in the acid zone will also increase the effort required to reach cleanup levels. These minerals will be soluble in the native groundwater, releasing their constituents to water flushed through the acid zone. The amount of time required to remove these minerals will be a function of the flow rate, the solubility of the mineral, and the amount of the mineral present in the aquifer. The iron and aluminum released to solution will likely precipitate as hydroxide minerals, adding to the acid conditions (Reactions 12-20 and 12-21); however, they will also scavenge trace metals from solution by adsorption and enhance remediation by this process. Gypsum may produce the greatest hindrance to meeting cleanup goals because its solubility is great enough that it provides sulfate to groundwater well above the drinking water standard, but its solubility is not high enough that it can be easily removed from the aquifer. Furthermore, the combination of high sulfate in the acid source water and calcium released from the dissolution of calcite, dolomite, and other calcium minerals can produce large quantities of gypsum in the pore spaces of the aquifer.

The funnel-and-gate system can be used to intercept and treat an acidic plume *in situ*. Appropriate neutralizing compounds such as carbonate minerals can be added to the reaction gate material to halt the spread of the acid front. A balance must be determined between the flux of acidity through the gate, the amount of neutralizing agent added, and plugging of the gate by mineral precipitation. For a highly concentrated plume of strong acidity and large size, it is likely that the gate material will need to be replaced during the life of the remediation system. Just as natural neutralization is effective for pH and metals but not for sulfate, the removal of this anion by neutralization in the reaction gate is not expected to achieve drinking water standards. Blowes and others[35] found in laboratory tests that sulfate and metals can be effectively reduced in concentration by a combination of neutralizing and reducing agents that could be incorporated in the reaction gate material. Their active agents consisted of a mixture of limestone for neutralization, solid-phase organic carbon to promote sulfate reduction and sediments as a source of bacteria. At a small-scale field test of a reaction wall held at a mine tailings site near Sudbury, Ontario, sulfate was reduced from 3,500 to 7 mg/L and iron was reduced from 1,000 to <5 mg/L.

REFERENCES

1. Singer, P.C. and Stumm, W., Acid mine drainage: the rate-determining step, *Science,* 167, 1121–1123, 1970.
2. Moses, C.O., Nordstrom, D.K., Herman, J.S., and Mills, A.L., Aqueous pyrite oxidation by dissolved oxygen and by ferric iron, *Geochim. Cosmochim. Acta,* 51, 1561–1571, 1987.
3. Moses, C.O. and Herman, J.S., Pyrite oxidation at circumneutral pH, *Geochim. Cosmochim. Acta,* 55, 471–482, 1991.
4. Kleinmann, R.L.P., Crerar, D.A., and Pacelli, R.R., Biogeochemistry of acid mine drainage and a method to control acid formation, *Mining Eng.,* 300–305, March 1981.
5. Kleinmann, R.L.P. and Crerar, D.A., Thiobacillus ferrooxidans and the formation of acidity in simulated coal mine environments, *Geomicrobiol. J.,* 1, 373–388, 1979.

6. Johnson, N.M., in *Geological Aspects of Acid Deposition*, Bricker, O.P., Eds., Butterworth, Boston, 1984, pp. 37–53.

7. Webb, J.A. and Sasowsky, I.D., The interaction of acid mine drainage with a carbonate terrane: evidence from the Obey River, north-central Tennessee, *J. Hydrol.*, 161, 327–346, 1994.

8. Litaor, M.I. and Thurman, E.M., Acid neutralizing processes in an alpine watershed Front Range, Colorado, U.S.A. I. Buffering capacity of dissolved organic carbon in soil solutions, *Appl. Geochem.*, 3, 645–652, 1988.

9. Moss, P.D. and Edmunds, W.M., Processes controlling acid attenuation in the unsaturated zone of a Triassic sandstone aquifer, (U.K.), in the absence of carbonate minerals, *Appl. Geochem.*, 7, 573–583, 1992.

10. Herlihy, A.T., Mills, A.L., and Herman, J.S., Distribution of reduced inorganic sulfur compounds in lake sediments receiving acid mine drainage, *Appl. Geochem.*, 3, 333–344, 1988.

11. Carignan, R. and Tessier, A., The co-diagenesis of sulfur and iron in acid lake sediments of southwestern Quebec, *Geochim. Cosmochim. Acta*, 52, 1179–1188, 1988.

12. Wicks, C.M., Herman, J.S., and Mills, A.L., Early diagenesis of sulfur in the sediments of lakes that receive acid mine drainage, *Appl. Geochem.*, 6, 213–224, 1991.

13. Moran, R.E. and Wentz, D.A., *Effects of Metal-Mine Drainage on Water Quality in Selected Areas of Colorado, 1972–73*, Colorado Water Conservation Board, Denver, CO 1974.

14. King, T.V.V., *Environmental Considerations of Active and Abandoned Mine Lands*, U. S. Geological Survey, Denver, Colorado, 1995.

15. Nordstrom, D.K., Jenne, E.A., and Ball, J.W., Redox Equilibria of Iron in Acid Mine Waters In *Chemical Modeling in Aqueous Systems*, Jenne, E.A., Eds., American Chemical Society, Washington, DC 1979.

16. Filipek, L.H., Nordstrom, D.K., and Ficklin, W.H., Interaction of acid mine drainage with waters and sediments of West Squaw Creek in the West Shasta Mining District, California, *Environ. Sci. Technol.*, 21, 388–396, 1987.

17. Karathanasis, A.D., Evangelou, V.P., and Thompson, Y.L., Aluminum and iron equilibria in soil solutions and surface waters of acid mine watersheds, *J. Environ. Qual.*, 17, 534–543, 1988.

18. Stollenwerk, K.G., Geochemical interactions between constituents in acidic groundwater and alluvium in an aquifer near Globe, Arizona, *Appl. Geochem.*, 9, 353–369, 1994.

19. Toran, L., Sulfate contamination in groundwater from a carbonate-hosted mine, *J. Contam. Hydrol.*, 2, 1–29, 1987.

20. Elberling, B., Nicholson, R.V., and Scharer, J.M., A combined kinetic and diffusion model for pyrite oxidation in tailings: a change in controls with time, *J. Hydrol.*, 157, 47–60, 1994.

21. Blowes, D.W. and Jambor, J.L., The pore-water geochemistry and the mineralogy of the vadose zone of sulfide tailings, Waite Amulet, Quebec, Canada, *Appl. Geochem.*, 5, 327–346, 1990.

22. Blowes, D.W., Reardon, E.J., Jambor, J.L., and Cherry, J.A., The formation and potential importance of cemented layers in inactive sulfide mine tailings, *Geochim. Cosmochim. Acta*, 55, 965–978, 1991.

23. Davis, E.C. and Boegly, W.J., A review of water quality issues associated with coal storage, *J. Environ. Qual.*, 10, 127–133, 1981.

24. Helz, G.R., Dai, J.H., Kijak, P.J., Fendinger, N.J., and Radway, J.C., Processes controlling the composition of acid sulfate solutions evolved from coal, *Appl. Geochem.*, 2, 427–436, 1987.

25. Peterson, S.R., Erikson, R.L., and Gee, G.W., *The Long Term Stability of Earthen Materials in Contact with Acidic Tailings Solutions*, U.S. Nuclear Regulatory Commission, Silver Spring, MD 1982.

26. Serne, R.J., Peterson, S.R., and Gee, G.W., *Laboratory Measurements of Contaminant Attenuation of Uranium Mill Tailings Leachates by Sediments and Clay Liners*, U.S. Nuclear Regulatory Commission, Silver Spring, MD 1983.

27. Peterson, S.R., Martin, W.J., and Serne, R.J., *Predictive Geochemical Modeling of Contaminant Concentrations in Laboratory Columns and in Plumes Migrating from Uranium Mill Tailings Waste Impoundments*, U.S. Nuclear Regulatory Commission, Silver Spring, MD 1986.

28. Davis, A. and Runnells, D.D., Geochemical interactions between acidic tailings fluid and bedrock: use of the computer model MINTEQ, *Appl. Geochem.*, 2, 231–241, 1987.

29. Miller, G.C., Lyons, W.B., and Davis, A., Understanding the water quality of pit lakes, *Environ. Sci. Technol.*, 30, 118A–123A, 1996.

30. Davis, A. and Ashenberg, D., The aqueous geochemistry of the Berkeley Pit, Butte, Montana, U.S.A., *Appl. Geochem.*, 4, 23–36, 1989.
31. Davis, A. and Galloway, J.N., Distribution of Pb between sediments and pore water in Woods Lake, Adirondack State Park, New York, U.S.A., *Appl. Geochem.*, 8, 51–65, 1993.
32. Jansson, M. and Ivarsson, H., Causes of acidity in the River Lillån in the coastal zone of central northern Sweden, *J. Hydrol.*, 160, 71–87, 1994.
33. Davis, A.O., Chapell, R., and Olsen, R.L., *The Use and Abuse of Eh Measurements: are They Meaningful in Natural Waters?* National Water Well Association, Dublin, OH, 1988.
34. Nicholson, R.V., Gillham, R.W., and Reardon, E.J., Pyrite oxidation in carbonate-buffered solution. II. Rate control by oxide coatings, *Geochim. Cosmochim. Acta*, 54, 395–402, 1990.
35. Blowes, D.W., Ptacek, C.J., Waybrant, K.R., Bain, J.D., and Robertson, W.D., *In Situ Treatment of Mine Drainage Water Using Porous Reactive Walls*, Environment and Energy Conference of Ontario on the "New Economy" Green Needs and Opportunitites Toronto, ON, 1994.

13 GEOCHEMISTRY OF ORGANIC COMPOUND CONTAMINATION

Natural and man-made organic compounds present a potential risk to the environment because of their toxicity (e.g., pesticides and herbicides), carcinogenicity (e.g., benzene and vinyl chloride) and the effects of reactions with the natural system, such as oxygen consumption. The focus of this chapter is on the geochemical interactions between organic contaminants and the natural environment. These reactions limit the mobility of some organic compounds by active biological/chemical degradation and adsorption, although decomposition of one compound may produce a more mobile and toxic daughter species. Degradation of organic compounds also may modify the environment such that inorganic species become more mobile and add to the overall contamination of the system.

Representative situations and types of organic contaminants are described in this chapter. Sites contaminated with petroleum hydrocarbons and chlorinated solvents are discussed in detail. General guidelines for sampling and modeling these sites as well as contaminant mobility and restoration are discussed. The landfill environment (Chapter 10) shares similarities with these types of organic contamination because landfills commonly contain these compounds.

13.1. REPRESENTATIVE TYPES OF ORGANIC CONTAMINATION

Petroleum Hydrocarbons

Crude oil and byproduct fuels, such as gasoline, diesel, and fuel oil, represent a common source of organic contamination in the environment because of the large volumes of these liquids used in commerce and the presence of water-soluble components in these materials. The more toxic, water-soluble, and mobile components of petroleum hydrocarbons are benzene, toluene, ethylbenzene and the three xylene isomers (ortho-, meta- and para-); therefore this discussion will highlight the BTEX compounds.

One of the best documented sites that includes a description of the geochemistry at a petroleum-contaminated location is the crude oil pipeline break near Bemidji, Minnesota. A rupture of the pipeline in 1979 contaminated the soil and shallow aquifer by oil spray and a floating oil body on the water table,[1] Figure 13-1. The total dissolved organic carbon (TDOC) concentration in the native groundwater upgradient of the contamination is approximately 2 to 3 mg/L, whereas beneath the oil spray area it was measured at 16 mg/L and in the vicinity of the oil body the TDOC concentration was 48 mg/L in 1987.[2] The increase in TDOC in groundwater from the oil spray area is almost totally non-volatile hydrocarbons, principally alkanes in the higher molecular weight series C_{15} to C_{28}. It is to be expected that the volatile organic compounds would be missing in the oil spray zone because they would volatilize and/or biodegrade during transport by recharge water through the vadose zone to the water table. In the groundwater directly affected by the oil body on the water table, only 58% of the TDOC is comprised of non-volatile hydrocarbons with the majority of the remainder being aromatic hydrocarbons (principally BTEX, higher alkylated benzenes and polycyclic aromatic hydrocarbons such as naphthalene). Benzene is the dominant individual volatile compound near the oil (70% of the volatile fraction) followed by the

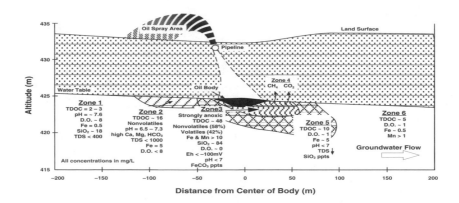

Figure 13-1 Bemidji (Minnesota) pipeline break. (TDOC: Total dissolved organic carbon; all concentrations in mg/L.)

C_{1-4} alkylated benzenes (14%). The persistence of hydrocarbons in the groundwater is quite variable. Toluene and o-xylene decrease to below detection limits within 10 m of the source, whereas benzene and ethylbenzene travel further downgradient. Eaganhouse et al.[2] found the aromatic hydrocarbon 1,2,3,4-tetramethylbenzene to be one of the most abundant and stable of the soluble oil constituents, and recommended its use as a marker for this plume. Apparent biodegradation products of the petroleum hydrocarbons in the anoxic portion of the plume near the oil body include phenol and a variety of organic acids with the principal compounds being acetic acid, cyclohexanoic acid and alkylated homologs of cyclohexanoic acid.[3]

The TDOC remains elevated (>10 mg/L) for a distance of about 75–80 m downgradient from the center of the oil body, and then rapidly declines. At a distance of 200 m from the source area, the evidence of contamination is nearly undetectable. Because a conservative contaminant would have traveled at least 500 m during the timeframe from the spill to the sampling event in 1987, Eaganhouse et al.[2] conclude that natural biodegradation is effectively limiting the migration of petroleum hydrocarbons and their byproducts at this site.

The geochemical interactions between the organic contaminants, groundwater and aquifer solid phase material at the Bemidji site were established by evaluating trends in water composition and conducting laboratory microcosm experiments. In the most strongly anoxic zone that extends about 50 m downgradient of the oil source area (Figure 13-1), the principal reactions appear to be oxidation of organic compounds, precipitation of siderite ($FeCO_3$) and a ferroan calcite, dissolution of iron and manganese oxides and outgassing of CH_4 and CO_2.[4] In this zone, the pH is near neutral and probably buffered by inorganic carbon species. The Eh measured in the field was –0.145 mV, which compared favorably with the Eh calculated for the sulfate/sulfide (–0.165 mV) and bicarbonate/methane (–0.221 mV) couples but not with the Fe and N couples (Fe^{3+}/Fe^{2+}, $Fe(OH)_3/Fe^{2+}$ and NO_3^-/NH_4^+) which resulted in unrealistically high Eh values (>0 mV).

At the Bemidji site, the oxidation/reduction reactions leading to degradation of the organic contaminants initially used and consumed the available dissolved oxygen in groundwater as the electron acceptor. The consumption of dissolved oxygen created the anoxic zone near the source area. Within this zone the primary available electron acceptors became

Mn(IV) in the solid MnO_2 and Fe(III) present in the aquifer as $Fe(OH)_3$. The oxidation/reduction reactions involving metals are likely mediated by naturally-occurring microorganisms. Using contaminated sediment from the Bemidji site, Lovley and others[5] showed that the bacterium GS-15 can anaerobically oxidize benzoate, toluene, phenol or p-cresol with Fe(III) as the sole electron acceptor. Dissolution of the metal oxide minerals provided for continued oxidation of the organic compounds and increases in dissolved levels of Mn and Fe to concentrations greater than 10 mg/L. In some areas of the anoxic zone near the contaminant source, the Mn and Fe solids have been depleted and degradation of the organics has occurred by methanogenesis. Temporal data show that the concentrations of Fe, Mn and CH_4 have varied significantly over time and that part of the plume has become more reducing over a 5-year period.

In addition to the effect of oxidation/reduction reactions on the concentrations of Mn and Fe, the presence of organic acids as intermediate degradation products has apparently enhanced the weathering of silica minerals at the Bemidji site. The SiO_2 concentration in groundwater increases from a background level of 18 mg/L to 84 mg/L in the anoxic zone. This is attributed to silicon complexing with organic ligands thereby increasing the solubility of the silicate minerals, primarily feldspar and quartz.[1] In an in situ microcosm experiment, Hiebert and Bennett[6] found localized mineral etching associated with surface-adhering bacteria at the Bemidji site. They postulate that the bacteria create a localized zone of intense weathering facilitated by the metabolic byproducts of biodegradation of the organic contaminants. Enhanced weathering increases the dissolved concentrations of the mineral components. The dissolved aluminum concentration did not increase as found for silicon because of the precipitation of a relatively insoluble aluminum solid that limited the aluminum concentration in groundwater.

In the aquifer downgradient of the anoxic zone, the groundwater becomes increasing more oxidizing resulting in the rapid degradation of remaining organic contaminants and the removal of Mn and Fe from solution to low concentrations in equilibrium with oxides of these metals. Organic degradation products (such as phenol and organic acids) that were observed in the anoxic zone were not detected in the oxic region probably because they are rapidly metabolized to CO_2 and H_2O and are not persistent in oxygenated water. As the organic ligands are consumed, the solubility of silicon minerals decreases and the level of dissolved silica decreases to background concentrations.[1] Examination of sand grains showed the presence of amorphous silica precipitates.

Other sites with petroleum contamination from spills and leaks show similar patterns in groundwater contamination. Near the source, the readily oxidizable organic matter consumes available oxygen and other electron acceptors in the aquifer. Some of the organic contaminants are degraded by this process, but a localized anaerobic zone is created that tends to mobilize metals, particularly Fe and Mn, in the aquifer. Methanogenic conditions may be produced near the source if the oxidizing capacity of the aquifer cannot maintain more oxidizing conditions. Downgradient of the source and the anaerobic zone, biodegradation becomes more efficient as aerobic bacteria become dominant. Aerobic bacteria more rapidly degrade the non-chlorinated organic compounds. Also, as the redox potential increases, the metal oxide minerals become oversaturated and Fe and Mn reprecipitate lowering their concentrations in groundwater. As a consequence, aquifers have the ability to naturally restore water quality impacted by petroleum contamination. The capability of a particular aquifer to restore itself depends on the degree of contamination and the aquifer's restoration capacity.

Chlorinated Solvents

The chlorinated solvents primarily consist of the aliphatic compounds tetrachloroethene (PCE), trichloroethene (TCE) and 1,1,1-trichloroethane (TCA). This type of solvent has been in use by the metal processing, electronics, dry cleaning, paint, pulp and paper, and textile manufacturing industries for more than 50 years. Spills and improper waste disposal methods have led to soil and aquifer contamination by these compounds and their degradation products. Under anaerobic conditions in groundwater, it has been shown that PCE and TCE can undergo dechlorination reactions in which chloride is lost from the compound forming the daughter products 1,1-dichloroethene (1,1-DCE), *cis*-1,2-dichloroethene (c-DCE), *trans*-1,2-dichloroethene (t-DCE), and vinyl chloride (VC).[7] The dehalogenation reaction of TCA also forms 1,1-DCE. Aerobic degradation of the chlorinated solvents does not appear to be as effective as anaerobic degradation, although the initial degradation products may themselves be more easily degraded in an aerobic environment.[8] Under oxidizing conditions, the chlorinated degradation products may be oxidized and hydrolyzed producing intermediate organic compounds such as dichloroacetic acid ($CHCl_2COOH$) and formate (CHOOH) with the ultimate stable products being CO_2, H_2O and Cl^-.[9] The final part of this process is termed mineralization because the organic compounds have been degraded to their inorganic compounds, which may then be incorporated into mineral phases.

The chlorinated solvents are dense nonaqueous phase liquids (DNAPLs) that present particularly challenging site characterization and remediation problems when they are present in the environment as pure product. The specific gravities of the solvents and some of their degradation products are listed in Table 13-1. As it moves through the unsaturated zone, the DNAPL spreads laterally, adheres to organic matter and minerals, and partially volatilizes.[10] Because the liquid is denser than water, it will pass through the water table, dissolve in the groundwater and accumulate at the base of the aquifer or less permeable units within the aquifer (Figure 13-2). The residual liquid left in the unsaturated zone will continue to volatilize and begin to degrade producing daughter products. The gasses are denser than air and will migrate laterally and vertically downward in the void spaces to the top of the water table. The gases dissolve into percolating soil water and into groundwater. Contamination of water by the solvent gases will likely occur in the aquifer both upgradient and downgradient of the initial spill/discharge point because the gases move in all directions from the source in response to vapor diffusion. The pool of DNAPL in the aquifer below the water table may be difficult to locate because the separate, dense liquid phase will move in response to gravity and not the hydraulic gradient. It may move up, down or across the hydraulic gradient and accumulate in depressions on the surface of low permeability aquifer material. The DNAPL will provide a continuing source of contamination to the groundwater until it completely dissolves.

The solubilities of the chlorinated solvents and their degradation products are given in Table 13-1 along with the U.S. EPA maximum contaminant level. Based on a solubility of 1,100 mg/L for TCE, it would require 277,000 liters of water to dissolve all the TCE in one

TABLE 13-1.

PHYSICAL PROPERTIES AND MAXIMUM CONTAMINANT LEVELS FOR CHLORINATED SOLVENTS AND DEGRADATION PRODUCTS

Compound	Specific gravity	Solubility (20°C) (mg/L)	MCL (mg/L)
PCE	1.631	150	0.005
TCE	1.466	1100	0.005
1,1,1-TCA	1.346	1500	0.2
1,1-DCE	1.25	400–2,500	0.007
1,2-DCE	1.27	600–6,300	0.07(*cis*); 0.1(*trans*)
VC	0.908	1.1–60	0.002

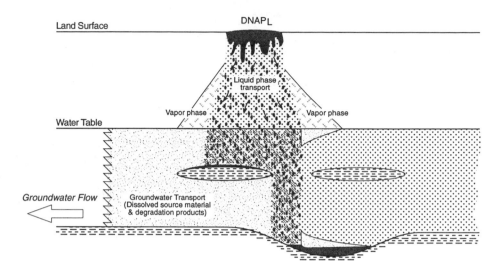

Figure 13.2 DNAPL migration in the subsurface (after Feenstra and Cherry[37]).

55 gallon (208 L) drum full of this solvent if the water was fully saturated with TCE. At an MCL of 0.005 mg/L, 55 gallons of TCE can contaminate 61 billion liters (16 billion gallons) of water at the MCL. Sixty-one billion liters of water would fill about 5 km² of an aquifer that is 10 m thick and has a porosity of 0.25. It is apparent from these values that a relatively small mount of a chlorinated solvent can impact very large quantities of groundwater and affect a wide area of an aquifer.

The concentrations of the solvents and their degradation products in contaminated groundwater at municipal landfills are typically in the low microgram/L range. This is probably due to the relatively small amounts of these products disposed in the landfill and the strong biodegradation and sorption reactions of the solvents in the landfill environment. At current and former military and industrial sites in Canada and the U.S. where much larger quantities of these materials have been used and disposed, the levels of TCE and TCA in groundwater have been reported in the hundreds to hundreds of thousands of micrograms/L.[11]

Concentrations of chlorinated solvents in groundwater greater than a few mg/L may be sufficient to create anaerobic conditions conducive to the reductive dechlorination of PCE and TCE as long as the solvents are not toxic to the microorganisms. At concentrations less than one mg/L it is unlikely that the solvents themselves can generate the necessary reducing conditions and they will be relatively stable and mobile. If the chlorinated solvents are disposed with other reactive organic matter, such as in sewage effluent, the oxidation of the other organic material may produce reducing conditions and degradation of PCE and TCE. At the Otis Air Force Base (Massachusetts) Barber et al.[12] found that reducing conditions were established near the disposal point of sewage effluent because of the high concentration of reactive organic matter, but that further downgradient the system remained oxidizing. Although some biodegradation occurs in the limited anaerobic zone near the effluent discharge location, the chlorinated solvents PCE, TCE and DCE and other organic compounds (such as dichlorobenzene and the isomers of nonylphenol) are relatively mobile and stable having persisted for more than 30 years in the aquifer.

13.2. SAMPLING CONSIDERATIONS

In evaluating the occurrence and mobility of organic compounds, the sampling program must include several provisions not commonly found in investigations for inorganic contaminants. Many of the organic contaminants are volatile and may outgas from water in the vadose zone and groundwater at the water table. If the organic contaminant is disposed as a separate organic liquid phase, it will also partially volatilize in the vadose zone. The organic gases will move away from the source and may be lost to the atmosphere or contaminate water in the vadose zone and the aquifer in a zone radiating from the source.[10] A light non-aqueous phase liquid (LNAPL) will accumulate on the water table and a DNAPL will accumulate on a bed of less permeable material within or at the base of the aquifer. Reinhard et al.[13] found increasing concentrations of chlorinated solvents including TCE, PCE and 1,1,1-TCA dissolved in groundwater at the bottom of the upper aquifer, underneath the chloride plume at the North Bay and Woolwich landfills (Ontario). They attributed this occurrence to the possible presence of a separate liquid phase of the solvent at the base of the aquifer. Sampling procedures must include methods of detecting the presence of these nonaqueous phase liquids, identifying their composition and measuring their extent. They represent a continuing source of contamination to an aquifer similar to a soluble mineral containing an inorganic contaminant.

Most organic contaminants are susceptible to chemical and or biological degradation under certain conditions. The petroleum hydrocarbons are typically most degradable under oxidizing conditions and the chlorinated solvents (PCE and TCE) are known to be degradable under reducing conditions, although they may be cometabolized and degraded under oxidizing conditions with a suitable substrate. (See Section 13.5.) The redox status of the aquifer should be determined by measuring the Eh of the groundwater. The concentrations of oxidants (e.g., O_2 and NO_3^-) and organic substrate (dissolved organic matter) should also be determined. It is also important to identity the dissolved organic matter because, if it is resistant to bacterial degradation, it may not stimulate bacterial growth and consequent breakdown of the organic contaminants. Barber et al.[12] found at the Otis Air Force Base that, although the dissolved organic matter was relatively high at about 4 mg/L, it was predominantly surfactants that were resistant to bacterial attack and did not enhance the removal of the trace organic contaminants. The efficiency of bacterial degradation may also be pH dependent, therefore an accurate measurement of this parameter also is necessary.

The movement of organic contaminants may be retarded by adsorption onto naturally-occurring organic solids in the vadose zone and the aquifer. The amount of organic carbon present in the solid phase is usually reported as a fraction of the total solids present (f_{oc}). This number is used in the calculation of the distribution coefficient (K_{oc}) and the retardation factor (R) for the individual organic compounds. (See Section 3.1.)

Gillham et al.[14] describe an in situ column procedure for measuring retardation factors for organic compounds. They tested the use of a column apparatus installed in advance of the cutting head of hollow-stem augers in which contaminants can be injected and breakthrough patterns recorded. They measured retardation factors for 5 halogenated organic compounds (bromoform, carbon tetrachloride, tetrachloroethylene, 1-2-dichlorobenzene, hexachloroethane) and found the results to be similar to the laboratory batch method and the correlation method with K_{ow}. The method has advantages over large scale tracer tests which are expensive and time-consuming and laboratory bench measurements in which disturbed material is commonly used. However, the *in situ* method is limited to a medium in which hollow-stem augers can be used to install the column and the hydraulic conductivity must be high enough (greater than 10^{-4} to 10^{-5} cm/s) to generate sufficient flow through the column. In addition, the Kd of the compound should probably not be greater than about 10 to 15 L/Kg to ensure breakthrough in a reasonable period of time.

At sites heavily contaminated with organic compounds a significant portion of the groundwater alkalinity may be due to organic acids. If this is the case, the concentrations of these organic acids must be determined and subtracted from the total alkalinity before dissolved inorganic carbon concentration is calculated from alkalinity.

13.3. GEOCHEMICAL MODELING

The more commonly used geochemical modeling codes (WATEQ2, MINTEQ and PHREEQE) have limited capability to simulate interactions of organic compounds with the aquifer system. This is primarily due to the lack of kinetic capabilities in these codes and the importance of reaction rates in the degradation of organic compounds. The MINTEQ database does have complexation constants for many organic compounds with metals, and can be used to model this type of interaction which enhances the mobility of metals and may inhibit degradation of the complexed organic compound.

For the Bemidji petroleum-contaminated site described in Section 13.1, Baedecker et al.[4] and Bennett et al.[1] used the NETPATH and PHREEQE codes to evaluate mass transfer and reaction progress along the flowpath. In lieu of modeling oxidation/degradation of individual organic compounds, the change in concentration of total dissolved carbon was simulated by allowing it to form carbonate minerals, methane and/or carbon dioxide gas. The modeling aided in the identification of the principle solid phases dissolving into or precipitating from the groundwater to account for changes in concentration and showed that large amounts of Fe are part of the cycle in the anoxic zone.

Kinetic models are usually developed to understand and simulate the reactions of organic compounds in the environment. For example, Semprini and McCarty[15] developed a kinetic model for the cometabolic transformation of methane and chlorinated solvents at a pilot-scale remediation effort at Moffett Field (California). The model included the processes of microbial growth, utilization of the electron donor (methane) and acceptor (oxygen), and the transformation of PCE and TCE to DCE and VC. The transport processes of advection, dispersion and adsorption were also included. The model was also used successfully to simulate the injection of phenol as the substrate for cometabolic decomposition of chlorinated solvents at Moffett Field.[16]

13.4. CONTAMINANT TRANSPORT

The chemical and biochemical processes occurring in the subsurface that impact the movement of organic contaminants are primarily adsorption/desorption and degradation. As described in detail in Chapter 3, adsorption/desorption is represented by a distribution coefficient (Kd) for the contaminant that relates adsorbed concentration (C_{ads}) to dissolved concentration (C_{aq}):

$$Kd = C_{ads}/C_{aq} = K_{oc}f_{oc}$$

The distribution coefficient of a contaminant at a particular site is a function of the attraction of the solid phase organic matter in the aquifer for the contaminant (as represented by the organic carbon partition coefficient, K_{oc}) and the fraction of organic carbon present in the aquifer solid phase (f_{oc}). The f_{oc} is measured as part of site characterization. The organic carbon partition coefficient can be calculated from the solubility of the compound in water or from the octanol/water partition coefficient (K_{ow}). K_{ow}s have been tabulated for most organic contaminants of concern.[17] In general, compounds with a relatively high water solubility such as alcohols, ketones, aldehydes, low molecular weight aliphatics (including the chlorinated solvents and their degradation products), benzene and toluene are not strongly adsorbed in typical aquifers and thus are mobile. Organic compounds of lower solubility including the xylenes, ethylbenzene, polynuclear aromatic hydrocarbons (such as naphthalene and phenanthrene) and PCBs are more strongly adsorbed onto organic solids in the aquifer and are consequently less mobile.

At Otis Air Force Base (Massachusetts) where secondary sewage effluent has been discharged to the subsurface for decades, Barber et al.[12] found that the relatively soluble chloroethenes and chlorobenzenes had the same general distribution as chloride suggesting very little adsorption or retardation, whereas the less soluble compounds, such as non-ylphenol and di-tert-butylbenzoquinone, appeared to have been retarded during ground-water transport by adsorption processes.

Variations in the mobility of organic compounds can lead to plume separation. For example, the K_{ow}s for TCE and its degradation products 1,1-DCE, 1,2-DCE and VC are 195, 5.4 to 30, 30, and 13, respectively.[11] The greater adsorption and retardation of TCE may result in a relatively small TCE plume compared to its degradation products which will be carried farther by the groundwater.

The other major process affecting the migration of organic compounds is degradation. Aerobic biodegradation of nonchlorinated organic compounds has been shown to be a relatively rapid process. During aerobic biodegradation, oxygen serves as the electron acceptor, as represented in the following case for benzene.

$$C_6H_6 + 7.5O_2 \rightarrow 6CO_2 + 3H_2O \tag{13-1}$$

In a field study conducted at the Borden (Ontario) site, Barker et al.[18] evaluated the rate of movement and persistence of 7.6 mg/L total BTX (benzene, toluene and xylene) added to the shallow unconfined sand aquifer. As a result of weak adsorption processes, the BTX components moved slightly slower through the aquifer than the conservative chloride tracer also added to the groundwater. After 434 days, essentially all of the added BTX had been biodegraded; with benzene being the only component to persist beyond 270 days. They found that BTX persisted longest in localized anoxic layers, and concluded that the dominant control over BTX biodegradation was the availability of dissolved oxygen. In related laboratory experiments, they found similar BTX degradation rates using native aquifer material, but essentially zero degradation using sterilized aquifer material as a control. This confirmed that degradation is catalyzed by microorganisms. Under anaerobic conditions in the laboratory with both active and sterile aquifer material, they found very little BTX degradation.

Nielsen and Christensen[19] studied the aerobic degradation of seven aromatic hydrocarbons (benzene, toluene, o-xylene, p-dichlorobenzene, o-dichlorobenzene, naphthalene and biphenyl). In laboratory batch experiments they contacted solutions with concentrations of these compounds in the range of 119 to 152 μg/L with eight samples of aquifer material collected over a small 15 m × 30 m section of an aerobic aquifer. For all of the tests, benzene, toluene, naphthalene and biphenyl were degraded to <2 μg/L within 1 month, whereas the degradation of o-xylene, p-dichlorobenzene and o-dichlorobenzene continued for a period of about 3 months until 5 to 20% of the compounds were left. There was a short lag time (maximum of 9 days) before degradation commenced, and a significant difference in the overall degradation rate for benzene, toluene, naphthalene and biphenyl dependent on locality (up to a factor of 15 for biphenyl). No degradation was measured in the sterile control microcosms.

In the presence of readily degradable organic contaminants such as aromatic hydrocarbons, the small amount of dissolved oxygen present in aerobic groundwater will be quickly consumed. For the complete oxidation of benzene shown in reaction 13-1, each mole of benzene requires 7.5 moles of dissolved molecular oxygen. Therefore, an initial dissolved oxygen concentration of 10 mg/L (3.13×10^{-4} mol/L) will only be able to oxidize 3.25 mg/L (4.17×10^{-5} mol/L) of benzene. Once the dissolved oxygen has been consumed near the contaminant source, the system will become anaerobic. For organic degradation to continue either other electron acceptors must be available in the aquifer to oxidize the organic contaminant or methanogenesis must occur. Typically, anaerobic biodegradation rates are much slower than aerobic biodegradation rates, and some aromatic compounds, such as benzene, have not been shown to degrade consistently under anaerobic conditions.[20]

Several naturally-occurring compounds can serve as electron acceptors for the oxidation of organic contaminants under anoxic aquifer conditions. These include nitrate, Mn(IV), Fe(III) and sulfate. In laboratory and field tests at the Borden (Ontario) site, Barbaro et al.[21] found that the addition of NO_3^- to the anaerobic system enhanced the removal of toluene, had no effect on benzene degradation, and decreased the removal of ethylbenzene and the xylene isomers. In a separate study of the biodegradation of benzene, toluene, and the isomers of xylene in anaerobic batch microcosms also using material from the Borden aquifer, Major et al.[22] found a greater degree of biodegradation. They found that for samples of water with 100 mg $NaNO_3$/L and initial concentrations of 3 mg/L each of benzene, toluene, o-xylene and m-xylene, after 62 days the remaining organic levels were 5, 2, 15, and 12%, respectively, of the initial concentrations.

Hutchins et al.[23] also evaluated the use of denitrification as an alternative to oxygen addition for the biorestoration of the jet fuel JP-4. In laboratory tests to which JP-4, nitrate and nutrients were added to uncontaminated aquifer material, toluene was relatively quickly removed over a 30–40 day period, whereas there was observed a 30 day lag time before biodegradation commenced for xylenes, ethylbenzene and 1,2,4-trimethylbenzene. Benzene was not significantly degraded over a period of 60 days. Identical tests with contaminated aquifer material showed much longer lag times and decreased rates of biodegradation. Benzene, ethylbenzene, and o-xylene were not significantly degraded over the 6-month time period for this experiment. Because of the variability in these results due to undetermined causes, they stressed the need for site-specific studies prior to reaching conclusions about the impact of denitrification on the degradation of organic contaminants.

As described above in Section 13.1, investigations at the Bemidji oil pipeline spill site confirmed the mechanism for oxidation and biodegradation of aromatic compounds with Fe(III) aquifer solids.[5] In addition, it appeared that Mn(IV) minerals also supplied oxidizing capacity to the aquifer.[1,4] However, because of the lower concentration of oxidized manganese minerals in this aquifer compared to iron oxides, the oxidizing capacity attributable to metal oxides at the Bemidji site was primarily due to the presence of Fe(III) oxides and oxyhydroxides. This is likely to be the case in most aquifer systems because of the predominance of oxidized iron weathering products over oxidized manganese minerals.

Under highly reducing conditions (Eh < –150 mV) where the reactive oxidized metal solids have been completely removed or were never present, sulfate may provide the necessary electron acceptor for organic degradation. Using material from a gasoline spill at Seal Beach (California) Haag et al.[24] showed in the laboratory that naturally-occurring sulfate was capable of degrading 60% of the toluene producing CO_2. The other six added substrates (benzene, ethylbenzene, p-xylene, o-xylene, 1,3,5-trimethylbenzene, and naphthalene) were not degraded when the toluene supply was constant, however, when toluene addition was stopped in the experiments, p-xylene degradation commenced. Because insignificant amounts of methane were formed, methanogenesis was excluded as a predominant transformation pathway, leaving sulfate as the most likely electron acceptor for this site. Furthermore, addition of sulfate stimulated the transformation of toluene in batch experiments.

If exogenous electron acceptors, such as O_2, NO_3^-, Fe^{3+}, and SO_4^{2-}, are not available in a system, then degradation of the organic compounds must occur through the processes of fermentation and methanogenesis.[8] Fermentative bacteria hydrolyze the organic contaminant polymers into simpler monomeric compounds such as alchohol, organic acids and acetate. The methane-producing bacteria, which can degrade only a limited range of compounds, utilize these simpler compounds for growth and energy. During methanogenesis, a simple monomeric compound, commonly acetate CH_3COO^-, with more than one carbon atom is both oxidized (to CO_2) and reduced (to CH_4) to complete the degradation process without a separate electron acceptor. Grbic–Galic[8] provides a comprehensive review of the fermentative/methanogenic degradation of chlorinated and nonchlorinated aromatic compounds and phenols by subsurface organisms.

Landfills are probably the most common organic-contaminated environments where methanogenic conditions develop. Wilson et al.[25] conducted laboratory microcosm experiments with aquifer material and leachate from a municipal landfill in Norman, Oklahoma. The primary objective was to evaluate the behavior of benzene, toluene, ethylbenzene, o-xylene, 1,1-dichloroethene, *trans*-1,2-dichloroethene, *cis*-1,2-dichloroethene, TCE and 1,2-dibromoethane under anaerobic, methanogenic conditions. They found that toluene, *cis*-1,2-dichloroethene and 1,2-dibromoethane degraded completely over a period of several months with a lag time for initiation of decomposition of a few weeks. The remaining compounds showed a longer lag time of several months and the need for close to a year or two of degradation before the compound disappeared from the test samples. Very little degradation was observed in the sterile experiments compared to the live microcosms.

Acton and Barker[26] also showed that aromatic hydrocarbons are degradable under the anaerobic conditions typically found at landfills. They found from field experiments conducted at the North Bay (Ontario) landfill the following order of biodegradability: toluene > m-xylene > ethylbenzene > 1,2,4-trimethylbenzene > o-xylene > benzene. Except for toluene, they found little degradation of these compounds during an in situ column experiment at the Borden (Ontario) site. They attributed the difference in degradation at the two sites to the ability of the indigenous microorganisms to acclimate to the contaminants. At the North Bay site, conditions were very reducing with active methanogenesis. The contaminants were degraded primarily by fermentation and methanogenesis. At the Borden site, conditions were less reducing and not methanogenic. Microbial activity at Borden was not as high as at the North Bay site, consequently degradation of the added contaminants was reduced.

At the site of an aviation gasoline spill that contaminated an aquifer in Traverse City, Michigan, Wilson et al.[27] showed that the monoaromatic hydrocarbons (BTEX) can be degraded under both aerobic and anaerobic/methanogenic conditions. In the anaerobic core of the plume, they showed high concentrations of methane and BTEX with no detectable oxygen. Surrounding the core was a second anaerobic zone with greatly reduced dissolved concentrations of BTEX, but still substantial concentrations of methane. In laboratory simulation experiments, they showed that the degradation of the BTEX compounds was accompanied by methane generation and found that the removal rate of BTEX under anaerobic conditions to be rapid and comparable to the rate under aerobic conditions. The majority of the BTEX compounds had disappeared from solution in the first 4 weeks of the experiments. Other studies have shown that the anaerobic biodegradation rate of the aromatic compounds is much slower than the aerobic rates, and benzene has not been shown to degrade consistently under anaerobic conditions.[20] Variations in site conditions and the amount of time the organisms have had to adapt to the contaminants and anaerobic condition may account for some of the differences in degradation rates.

Degradation of the chlorinated solvents appears to be facilitated by anaerobic conditions. Reinhard et al.[13] found PCE, TCE and 1,1,1-TCA in groundwater adjacent to the North Bay (Ontario) landfill, but not further downgradient, and attributed the disappearance to microbial dechlorination under methanogenic conditions. At the Woolwich (Ontario) landfill, which was also reducing but not apparently methanogenic, removal of the chlorinated compounds was less efficient.

The reductive dechlorination of PCE and TCE produces the daughter products c-DCE, t-DCE, VC and, eventually ethylene. However, VC is less degradable anaerobically than TCE and tends to accumulate. Vinyl chloride and the two forms of DCE are easier to degrade in aerobic than in anaerobic conditions.[11]

Under aerobic conditions, such as found at Otis Air Force Base (Massachusetts) downgradient of the disposal point for secondary sewage effluent, Barber and others[12] noted that the chlorinated solvents are relatively persistent in groundwater. They state that the presence of dissolved oxygen (0.1 to 5 mg/L) in the groundwater throughout much of the plume makes reductive dechlorination unlikely.

13.5. REMEDIATION

The properties of organic compounds provide some challenges and special opportunities for remediation that are not in common with inorganic contaminants. The physical properties of volatility and specific gravity can enhance, but, in general, will inhibit remediation. In the unsaturated zone, restoration can be enhanced by the removal of volatile petroleum hydrocarbons by vapor extraction systems, however the movement of volatile organic gases can also spread contamination away from the source and contaminate groundwater over a larger area. Light and dense nonaqueous phase liquid contaminants will also present a remediation challenge as they accumulate on the water table (LNAPLs) and below the water table (DNAPLs) complicating the process of locating and removing pure product sources of contamination. This section will focus on the chemical and biochemical aspects of the remediation of organic compounds that must be considered in designing cleanup methods and estimating cleanup times.

As described in the section on contaminant mobility (13.4), the movement of dissolved organic contaminants can be affected by adsorption/desorption processes that are a function of the affinity of the solid phase organic carbon in the aquifer for the contaminant and the concentration of solid adsorbent in the system. Compounds that are strongly attracted to the solid phase, such as polynuclear aromatic hydrocarbons (PAHs), may move through an aquifer at a rate that is orders of magnitude less than the groundwater flow rate, however, even compounds that are not strongly adsorbed, such as benzene and toluene, will move at a slower rate than groundwater. This retarded movement will add to the amount of time required to flush organic contaminants from the aquifer. One process that has been tested to mobilize DNAPLs in an aquifer and which should also enhance desorption of organic contaminants is to add a surfactant to the water in the aquifer. As described by Wunderlich et al.,[10] surfactants form strong dissolved complexes with the organic contaminants thereby increasing the solubility in water of the pure product. Aqueous complexation should also compete with adsorption for the contaminants, thereby increasing solution concentration and enhancing removal. Part of the selection criteria for the complexing surfactants are that they be nontoxic and biodegradable. At the Borden (Ontario) site, a field test showed that the addition of a 2% mixture of nonylphenol ethoxylate and a phosphate ester of an alkylphenol ethoxylate to the groundwater increased the solubility of PCE from 200 mg/L to greater than 10,000 mg/L, significantly increasing the efficiency of PCE removal.[10]

The removal of volatile organic compounds from an aquifer can also be enhanced by air sparging in which air is bubbled into the groundwater. As the air bubbles rise through the plume, the volatile compounds partition into the gas phase and move upward out of the aquifer into the vadose zone. The gas may be collected by a vapor extraction system. The air may be injected at a well screened below the zone of contamination or in a trench constructed across the flow path.[28] Well sparging systems may be appropriate for very localized zones of contamination, whereas the trench system can extend across the entire width of a plume. Schima et al.[29] showed that for a sparge well installed in a sandy aquifer in which the air injection point was about 12 feet below the water table, the radius of influence was 9 feet for continuous sparging over several hours. They also found that some of the air became trapped in the aquifer above the injection point, thereby providing a potential long-term source of oxygen to the groundwater. This should enhance the growth of aerobic bacteria.

The removal of organic contaminants from groundwater can also be enhanced by degradation of the compounds. Because of the generally slow rate of simple chemical oxidation and decomposition of organic compounds, these reactions are usually catalyzed by naturally occurring microorganisms. The growth environment of the bacteria that facilitate decomposition of the organic contaminant can be optimized during the remediation process. For example, petroleum hydrocarbons are most quickly degraded under aerobic conditions but they rapidly consume the available dissolved oxygen in groundwater. Consequently, adding molecular oxygen to the aquifer via air/O_2 sparging or addition of hydrogen peroxide (H_2O_2) will maintain the aerobic conditions conducive to the decomposition of petroleum hydrocarbons.

The relatively low solubility of oxygen in water (7–12 mg/L) requires continuous addition of oxygen to maintain oxidizing conditions. As described above in Section 13.4, other electron acceptors may serve the purpose of oxygen in the biodegradation process. Nitrate, in particular, has been proposed as an additive to enhance the remediation of gasoline-contaminated aquifers.[22] Nitrate minerals are very soluble in water and nitrate movement is not significantly retarded by adsorption reactions with the aquifer solids. Therefore, it should be possible to add sufficient nitrate to degrade all of the organic contaminants in a single application. Because of the potential toxicity of nitrate, the design of the remediation system would need to allow for almost complete denitrification on degradation of the organic contaminant, however any residual nitrate should be easier to remove from the aquifer than the original organic compounds. The ability of nitrate to degrade the BTEX compounds has not been shown consistently at all sites.[26] In particular, benzene may be persistent.[23]

In addition to the need to add electron acceptors to enhance biodegradation of nonchlorinated organic contaminants, microorganisms also require inorganic nutrients to grow, particularly nitrogen (N), potassium (K) and phosphorus (P). A lack of these nutrients may slow down or stop bioremediation. Armstrong et al.[30] found that the addition of N, K and P enhanced the degradation of toluene in contaminated groundwater, with K and P enhancing degradation by a factor of two. The addition of dissolved oxygen did not increase the removal rate of toluene suggesting that conditions were sufficiently aerobic. When oxygen is added to stimulate growth of aerobic bacteria and degradation of organic contaminants, it is more likely that one of the nutrients will control the growth rate. Allen[31] found that nitrate limited the growth of BTEX-degrading microorganisms when oxygen was readily available.

Chlorinated hydrocarbons are degraded under anaerobic conditions through the process of reductive dechlorination. However, the byproducts of degradation of PCE and TCE are the DCE isomers and VC, which are also groundwater contaminants. It has been shown that under aerobic conditions these chlorinated solvents and their daughter products can also be degraded through the process of cometabolism. In this process a substrate, such as methane, toluene or phenol, that degrades under oxidizing conditions is added to the zone of contamination along with oxygen to encourage bacterial growth. During metabolism involving the primary substrate, the trace level solvents and their byproducts are also consumed. At the Moffett Field (California) field test site, Hopkins and McCarty[7] found that the addition of 9 mg/L toluene or 12.5 mg/L phenol to groundwater resulted in removal efficiencies of greater than 90% for TCE, c-1,2-DCE and VC and about 74% for t-1,2-DCE. However, only 50% of the 1,1-DCE was transformed with phenol addition, and the presence of 1,1-DCE appeared to hinder the aerobic decomposition of TCE. Both hydrogen peroxide and pure oxygen performed adequately as the added electron acceptor. The phenol and toluene added to the groundwater at Moffett Field were also efficiently removed by the biodegradation process. At the adjacent monitoring well the phenol concentration never exceeded 2 μg/L and toluene did not exceed its detection level of 1 μg/L. As mentioned in Section 13.3, the degradation of phenol and the chlorinated solvents at this site was successfully simulated with a kinetic reaction model developed by Semprini et al.[16]

The permeable reaction gate technology holds promise for in situ treatment and degradation of organic contaminants.[32] Addition of an oxygen source, such as a solid phase oxygen-releasing compound,[33] can create conditions conducive to the growth of aerobic bacteria and the consumption of most organic contaminants. The use of zero-valent iron to create strongly reducing conditions in the reaction zone holds promise for the chlorinated solvents that are more likely to be degraded under anaerobic conditions. Gillham and O'Hannesin[34] conducted laboratory tests with zero-valent iron to evaluate its effectiveness to degrade 14 chlorinated methanes, ethanes, and ethenes. They found that iron enhanced degradation rates for all compounds, except dichloromethane, by 3 to 13 orders of magnitude. The process was abiotic and the rates declined with decreasing degree of chlorination. The degradation of PCE and TCE resulted in the formation of c-DCE and VC at a level of about 10% of the parent concentration, however these byproducts were not persistent. The formation of secondary iron minerals, such as $FeCO_3$, may clog the pores of the reaction gate or coat the reactive metallic iron surface. These reactions should be considered in the design of this type of in situ remediation system.[35] Gillham[36] provides an excellent review of the metal-enhanced treatment technology and a discussion of several case studies.

REFERENCES

1. Bennett, P.C., Siegel, D.E., Baedecker, M.J., and Hult, M.F., Crude oil in a shallow sand and gravel aquifer — I. Hydrogeology and inorganic geochemistry, *Appl. Geochem.*, 8, 529–549, 1993.
2. Eganhouse, R.P., et al., Crude Oil in a Shallow Sand and Gravel Aquifer — II. Organic Geochemistry, *Appl. Geochem.*, 8, 551–567, 1993.
3. Cozzarelli, I.M., Eganhouse, R.P., and Baedecker, M.J., Transformation of Monoaromatic Hydrocarbons to Organic Acids in Anoxic Groundwater Environment, *Environ. Geol. Water Sci.*, 16, 135–141, 1990.
4. Baedecker, M.J., Cozzarelli, I.M., Eganhouse, R.P., Siegel, D.I., and Bennett, P.C., Crude oil in a shallow sand and gravel aquifer — III. Biogeochemical reactions and mass balance modeling in anoxic groundwater, *Appl. Geochem.*, 8, 569–586, 1993.
5. Lovley, D.R., et al., Oxidation of Aromatic Contaminants Coupled to Microbial Iron Reduction, *Nature*, 339, 297–300, 1989.
6. Hiebert, F.K. and Bennett, P.C., Microbial Control of Silicate Weathering in Organic-Rich Ground Water, *Science*, 258, 278–281, 1992.
7. Hopkins, G.D. and McCarty, P.L., Field Evaluation of In Situ Aerobic Cometabolism of Trichloroethylene and Three Dichloroethylene Isomers Using Phenol and Toluene as the Primary Substrates, *Environ. Sci. Technol.*, 29, 1628–1637, 1995.
8. Grbic–Galic, D., Methanogenic Transformation of Aromatic Hydrocarbons and Phenols in Groundwater Aquifers, *Geomicrobiol.*, 8, 167–200, 1990.
9. Little, C.D. et al., Trichloroethylene biodegradation by methane oxidizing bacteria, *Appl. Environ. Microbiol.*, 54, 951–956, 1988.
10. Wunderlich, R.W., Fountain, J.C., and Jackson, R.E., *In Situ* Remediation of Aquifers Contaminated with Dense Nonaqueous Phase Liquids by Chemically Enhanced Solubilization, *J. Soil Contam.*, 1, 361–378, 1992.
11. Oleszkiewicz, J.A. and Elektorowicz, M., Groundwater Contamination with Trichloroethylene: The Problem and Some Solutions — A Review, *J. Soil Cont.*, 2, 205–228, 1993.
12. Barber, L., B., II, Thurman, E.M., Schroeder, M.P., and LeBlanc, D.R., Long-term Fate of Organic Micropollutants in Sewage-Contaminated Groundwater, *Environ. Sci. Technol.*, 22, 205–211, 1988.
13. Reinhard, M., Goodman, N.L., and Barker, J.F., Occurrence and Distribution of Organic Chemicals in Two Landfill Leachate Plumes, *Environ. Sci. Technol.*, 18, 953–961, 1984.
14. Gillham, G.W., Robin, M.J.L., and Ptacek, C.J., A Device for In Situ Determination of Geochemical Transport Parameters. 1. Retardation, *Ground Water*, 28, 666–672, 1990.

15. Semprini, L. and McCarty, P.L., Comparison between Model Simulations and Field Results for *In Situ* Biorestoration of Chlorinated Aliphatics: Part 2. Cometabolic Transformations, *Ground Water,* 30, 37–44, 1992.

16. Semprini, L., Hopkins, G.D., and McCarty, P.L., in *Bioremediation of Chlorinated and Polycyclic Aromatic Hydrocarbon Compounds,* eds., Hinchee, R.E., Leeson, A., Semprini, L., and Ong, S.K., 248–254, Lewis Publishers, Boca Raton, 1994.

17. Nyer, E., Boettcher, G., and Morello, B., Using the Properties of Organic Compounds to Help Design a Treatment System, *GWMR,* 81–86, 1991.

18. Barker, J.F., Patrick, G.C., and Major, D., Natural Attenuation of Aromatic Hydrocarbons in a Shallow Sand Aquifer, *Groun Water Monit. Remed.,* 64–71, 1987.

19. Nielsen, P.H. and Christensen, T.H., Variability of biological degradation of aromatic hydrocarbons in an aerobic aquifer determined by laboratory batch experiments, *J. Contam. Hydrol.,* 15, 305–320, 1994.

20. McAllister, P.M. and Chiang, C.Y., A Practical Approach to Evaluating Natural Attenuation of Contaminants in Ground Water, *GWMR,* 161–173, 1994.

21. Barbaro, J.R., Barker, J.F., Lemon, L.A., and Mayfield, C.I., Biotransformation of BTEX under anaerobic, denitrifying conditions: Field and laboratory observations, *J. Contam. Hydrol.,* 11, 245–272, 1992.

22. Major, D.W., Mayfield, C.I., and Barker, J.F., Biotransformation of Benzene by Denitrification in Aquifer Sand, *Ground Water,* 26, 8–14, 1988.

23. Hutchins, S.R., Sewell, G.W., Kovacs, D.A., and Smith, G.A., Biodegradation of Aromatic Hydrocarbons by Aquifer Microorganisms under Denitrifying Conditions, *Environ. Sci. Technol.,* 25, 68–76, 1991.

24. Haag, F., Reinhard, M., and McCarty, P.L., Degradation of Toluene and p-Xylene in Anaerobic Microcosms: Evidence for Sulfate as a Terminal Electron Acceptor, *Envir. Toxic. Chem.,* 10, 1379–1389, 1991.

25. Wilson, B.H., Smith, G.B., and Rees, J.F., Biotransformations of Selected Alkylbenzenes and halogenated Aliphatic Hydrocarbons in Methanogenic Aquifer Material: A Microcosm Study, *Environ. Sci. Technol.,* 20, 997–1002, 1986.

26. Acton, D.W. and Barker, J.F., In Situ Biodegradation Potential of Aromatic Hydrocarbons in Anaerobic Groundwaters, *J. Contam. Hydrol.,* 9, 325–352, 1992.

27. Wilson, B.H., Wilson, J.T., Kampbell, D.H., Bledsoe, B.E., and Armstrong, J.M., Biotransformation of Monoaromatic and Chlorinated Hydrocarbons at an Aviation Gasoline Spill Site, *Geomicrobiol. J.,* 8, 225–240, 1990.

28. Pankow, J.F., Johnson, R.L., and Cherry, J.A., Air Sparging in Gate Wells in Cutoff Walls and Trenches for Control of Plumes of Volatile Organic Compounds, VOCs), *Ground Water,* 31, 654–663, 1993.

29. Schima, S., LaBrecque, D.J., and Lundegard, P.D., Monitoring Air Sparging Using Resistivity Tomography, *GWMR,* 16, 131, 1996.

30. Armstrong, A.Q., Hodson, R.E., Hwang, H.-M., and Lewis, D.L., Environmental Factors Affecting Toluene Degradation in Ground Water at a Hazardous Waste Site, *Environ. Toxicol. Chem.,* 10, 147–158, 1991.

31. Allen, R.M., *Fate and Transport of Dissolved Monoaromatic Hydrocarbons during Steady Infiltration through Soil,* University of Waterloo, 1991.

32. Smyth, D.J.A., Cherry, J.A., and Jowett, R.J., *Funnel-and-Gate for In Situ Groundwater Plume Containment,* Superfund XV Washington, DC, 1994.

33. Bianchi-Mosquera, G.C., Allen-King, R.M., and Mackay, D.M., Enhanced Degradation of Dissolved Benzene and Toluene Using a Solid Oxygen-Releasing Compound, *GWMR,* 120–128, 1994.

34. Gillham, R.W. and O'Hannesin, S.F., Enhanced Degradation of Halogenated Aliphatics by Zero-Valent Iron, *Ground Water,* 32, 958–967, 1994.

35. Tratnyek, P.G., Johnson, T., and Schattauer, A., *Interfacial Phenomena Affecting Contaminant Remediation with Zero-Valent Iron Metal,* 1–589–592, American Chemical Society, Atlanta, GA, 1995.

36. Gillham, R.W., in *Advances in Groundwater, Pollution Control and Remediation,* (ed., Aral, M.M.) 249–274, Kluwer Academic Publishers, Netherlands, 1996.

37. Feenstra, S. and Cherry, J.A. Subsurface contamination by dense nonaqueous phase liquid (DNAPL) chemicals. International Groundwater Symposium, Halifax, Nova Scotia, 1988.

A POTENTIAL GROUNDWATER CONCENTRATION-LIMITING SOLID PHASES

The accompanying table lists solid phases that have been reported to equilibrate with groundwater. At equilibrium these solid phases may limit the groundwater concentration for one of their constituents. (See Section 3.2 for a discussion of mineral equilibrium and its limitation on groundwater concentration.) The solubility of a mineral can vary over many orders of magnitude dependent on such factors as the pH, Eh, ionic strength, and dissolved concentrations of ligands for each particular system. Because of the variability of solubility, the concentration of the element limited by the solid must be determined for the site-specific conditions. This list of solid phases was primarily compiled from Rai and Zachara,[1] Lindsay,[2] Appelo and Postma,[3] Presser and Swain,[4] Baltpurvins et al.,[5] Palmer and Wittbrodt,[6] Hem and Lind,[7] and Santillan-Medrano and Jurinak.[8]

REFERENCES

1. Rai, D. and Zachara, J.M., *Chemical Attenuation Rates, Coefficients, and Constants in Leachate Migration a Critical Review,* Vol. 1, Electric Power Research Institute, Palo Alto, CA 1984.
2. Lindsay, W.L., *Chemical Equilibria in Soils,* John Wiley, New York, 1979.
3. Appelo, C.A.J. and Postma, D., *Geochemistry, Groundwater and Pollution,* A.A. Balkema, Rotterdam, 1994.
4. Presser, T.S. and Swain, W.C., Geochemical evidence for Se mobilization by the weathering of pyritic shale, San Joaquin Valley, California, U.S.A., *Appl. Geochem.,* 5, 703-717 1990.
5. Baltpurvins, K.A., Burns, R.C., Lawrance, G.A., and Stuart, A.D., Effect of pH and anion type of the aging of freshly precipitated iron(III) hydroxide sludges, *Environ. Sci. Technol.,* 30, 939-944, 1996.
6. Palmer, C.D. and Wittbrodt, P.R., Processes affecting the remediation of chromium-contaminated sites, *Env. Health Persp.,* 92, 25–40, 1991.
7. Hem, J.D. and Lind, C.J., Chemistry of manganese precipitation in Pinal Creek, Arizona, USA: A laboratory study, *Geochim. Cosmochim. Acta,* 58, 1601–1613, 1994.
8. Santillan-Medrano, J. and Jurinak, J.J., The chemistry of lead and cadmium in soil: solid phase formation, *Soil Sci. Soc. Amer. Proc.,* 39, 851–856, 1975.

Element	Potential groundwater concentration-limiting solids
Al	$Al(OH)_3$ am, gibbsite $(Al[OH]_3)$, clays — kaolinite $(Al_2Si_2O_5[OH]_4)$, halloysite $(Al_2O_3 \cdot 2SiO_2 \cdot 4H_2O)$
	Low pH, high sulfate: basalunite, alunite $(KAl_3[SO_4]_2[OH]_6)$, jurbanite $(AlOHSO_4)$
As	Oxidizing: scorodite $(FeAsO_4 \cdot 2H_2O)$, $Pb_3(AsO_4)_2$, $Mn_3(AsO_4)_2$
	Reducing: AsS_2
B	None
Ba	pH < 9: barite $(BaSO_4)$
	pH > 9: witherite $(BaCO_3)$
Be	$Be(OH)_2$
Ca^{2+}	Calcite $(CaCO_3)$, gypsum $(CaSO_4 \cdot 2H_2O)$
Cd	High pH: otavite $(CdCO_3)$, $CdOHSO_4$
	Near neural pH: $Cd_3(PO_4)_2$
	Reducing: greennokite: CdS
Cl^-	Halite $(NaCl)$ at very high (>200,000 ppm) conc.
CO_3^{2-}/HCO_3^-	pH > 6: calcite
	pH >≈ 7.5: rhodocrosite $(MnCO_3)$, otavite $(CdCO_3)$, cerrusite $(PbCO_3)$, witherite $(BaCO_3)$
	Reducing: Siderite $(FeCO_3)$
Cr	Oxidizing: Cr_2O_3, chromatite $(CaCrO_4)$, hashemite $(BaCrO_4)$, crocoite $(PbCrO_4)$, iranite $(PbCrO_4 \cdot H_2O)$
	Reducing: $Cr(OH)_3$ am, $(Fe,Cr)(OH)_3$, $FeCr_2O_4$, Cr_2O_3
Cu	Oxidizing: $CuFe_2O_4$ (cupric ferrite)
	Reducing: $Cu_2Fe_2O_4$ (cuprous ferrite), covellite (CuS)
F	Fluorite (CaF_2)
Fe	Oxidizing: ferrihydrite $(Fe[OH]_3)$, goethite $(\alpha\text{-}FeOOH)$, lepidocrocite $(\gamma\text{-}FeOOH)$
	Oxidizing, acidic: jarosite $(KFe_3[SO_4]_2[OH]_6)$, alunite $(KAl_3[SO_4]_2[OH]_6)$
	Reducing, alkaline: siderite $Fe(CO_3)$
	Reducing, sulfide present: amorphous ferrous sulfide (FeS), mackinawite (FeS), pyrite (FeS_2)
	Evaporite, at mine sites: melanterite $(FeSO_4 \cdot 7H_2O)$, copiapite $([Fe^{2+}Fe^{3+}]4[SO_4]_6[OH]_2 \cdot 20H_2O)$ rozenite $(FeSO_4 \cdot 4H_2O)$
Hg	Hg°, HgS
K^+	Illite
Mg^{2+}	pH < 7.5: none
	pH > 7.5: sepiolite $(Mg_4Si_6O_{15}[OH]_2 \cdot H_2O)$, possibly dolomite and clays (montmorillinite, chlorite)
Mn	Oxidizing: nsutite $(MnO_{1.9})$. birnessite $(\delta\text{-}MnO_2)$
	Reducing: manganite $(\gamma\text{-}MnOOH)$, Mn_2O_3, hausmannite (Mn_3O_4)
	Reducing, alkaline: rhodocrosite $(MnCO_3)$
Mo	Oxidizing: $Fe_2(MoO_4)_3$, wulfenite $(PbMoO_4)$, powellite $(PbMoO_4)$, $CaMoO_4$
	Reducing: molybdenite (MoS_2)
Na^+	pH < 9: none. pH > 9: possibly albite $(NaAlSi_3O_8)$
	Evaporative conditions: mirabilite $(Na_2SO_4 \cdot 10H_2O)$, bloedite $(MgNa_2[SO_4]_2 \cdot 4H_2O)$
Ni	Oxidizing: $NiFe_2O_4$, $Ni(OH)_2$
	Reducing: NiS
NO_3^-	Nitrate reduction to nitrogen, plant uptake
Pb	Cerrusite $(PbCO_3)$, anglesite $(PbSO_4)$, $Pb(OH)_2$, blixite $(Pb_4[OH]_6Cl_2)$, hydroxypyromorphite $(Pb_{10}[PO_4]_6[OH]_2)$, powellite $(PbMoO_4)$
PO_4^{3-}	Apatite $(Ca_3[PO_4]_2)$; variscite $(AlPO_4 \cdot 2H_2O)$, strengite $(FePO_4 \cdot 2H_2O)$, $MnHPO_4$
S	Oxidizing: gypsum $(CaSO_4 \cdot 2H_2O)$
	Reducing: S°, mackinawite (Fe/S), pyrite (FeS_2)
Si	Amorphous silica $(SiO_2 am)$, clays
U	Reducing: uraninite (UO_2)
Zn	Zincite (ZnO), hydrozincite $(Zn_4[OH]_6[CO_3]_2 \cdot xH_2O)$, franklinite $(ZnFe_2O_4)$

INDEX

multiple hypotheses, 83–84
use of saturation indices, 84
validation, 99
Mole, 6

N

Natural organic matter, 23
Natural restoration, 121–124, 203
Nernst Equation, 32, 94
Neutralization reactions, 14, 185–188
Nitrate, 7, 78

O

Organic acids, 147, 193, 203
Organic compound contamination, 201–203
  Bemidji, MN, 201–203
  chlorinated solvents, 203–205
  contaminant transport, 207–211
  gasoline spills, 128, 210
  kinetics, 207
  petroleum hydrocarbons, 201–203
  pH, 202–203
  modeling, 207
  remediation, 211–213
    aerobic degradation, 211–212
    anaerobic degradation, 212–213
    air sparging, 211–212
    cometabolism, 212
    funnel and gate, 213
    hydrogen peroxide, 212
    intrinsic (natural), 201–202, 208
    nutrients, 212
    oxygen-releasing compound, 213
    reaction lag time, 210
    vapor extraction systems, 211
    zero-valent iron, 213
  sampling, 206–207
Organic matter, 23–24, 149
  adsorption/desorption processes, 48–49,
    54–55, 79
  concentration, 148–149
  oxidation, 42, 153–154
Organic solute adsorption, 48–49
Ostwald ripening, 65
Otavite (CdCO$_3$), Appendix A
Oxidation-reduction processes, 29–32,
    122–123
Oxidizing Capacity, 15–16

P

Petroleum hydrocarbons, see LNAPLs and BTEX
pH, 10–12, 154
  buffering, 11–12
  dependent charge, 48–49

effect on mineral solubility, 70
Eh diagrams, 33–35, 184
measurement, 141–143
Phase rule, 25
Phases
  gas, 3
  solid, 3
  solution, 3, 5
Piper diagrams, see Trilinear diagrams
Plugging, 105–108
Pristine point of zero charge (PPZC), 57–58,
    136–137
Pump-and-treat, 129–132
  flush cycles, 130–132
  geochemical enhancements, 134–137
  tailing, 129–130
Pyrite (FeS$_2$), 79, 128, 133, 149, 158, 183, 189, 193

R

Reaction rates, 72–73, 79–81
Reactive minerals, 23, 65, 80–81, 116, 148–149,
    156, 191–192
Redox couples, 37, 147–148
Redox disequilibrium, 37, 80–81
Redox poising, 15, 137
Redox potential (Eh), 15–17, 29, 37
  effect on mineral solubility, 71
  at landfills, 153–155
  measuring, 143–144
  modeling, 103–105, 109–111
Redox processes, see oxidation-reduction
    processes
Redox-sensitive elements, 30–31, 36–37
Reducing Capacity, 15–17
Restoration/Remediation, 121–138
  acid sites, 197–198
  geochemical enhancements, 134–137
  landfills, 160–163
  modeling (simulating), 137–138
  natural/intrinsic, 121–124
  organics, 211–213
  unwanted side effects, 105–108, 128–129
  working toward equilibrium, 126–128
Retardation (R), 55
Rhodochrosite (MnCO3), 156
Rock
  composition, 21–22
  types, 21
Roll-front uranium deposits, 123–124

S

Sampling Programs, 141–149, 157–158
  alkalinity, 145–146
  Eh, 143–144
  electrical balance, 7–9